IMAGES
of America

RALEIGH *and* WAKE COUNTY FIREFIGHTING

Volume II

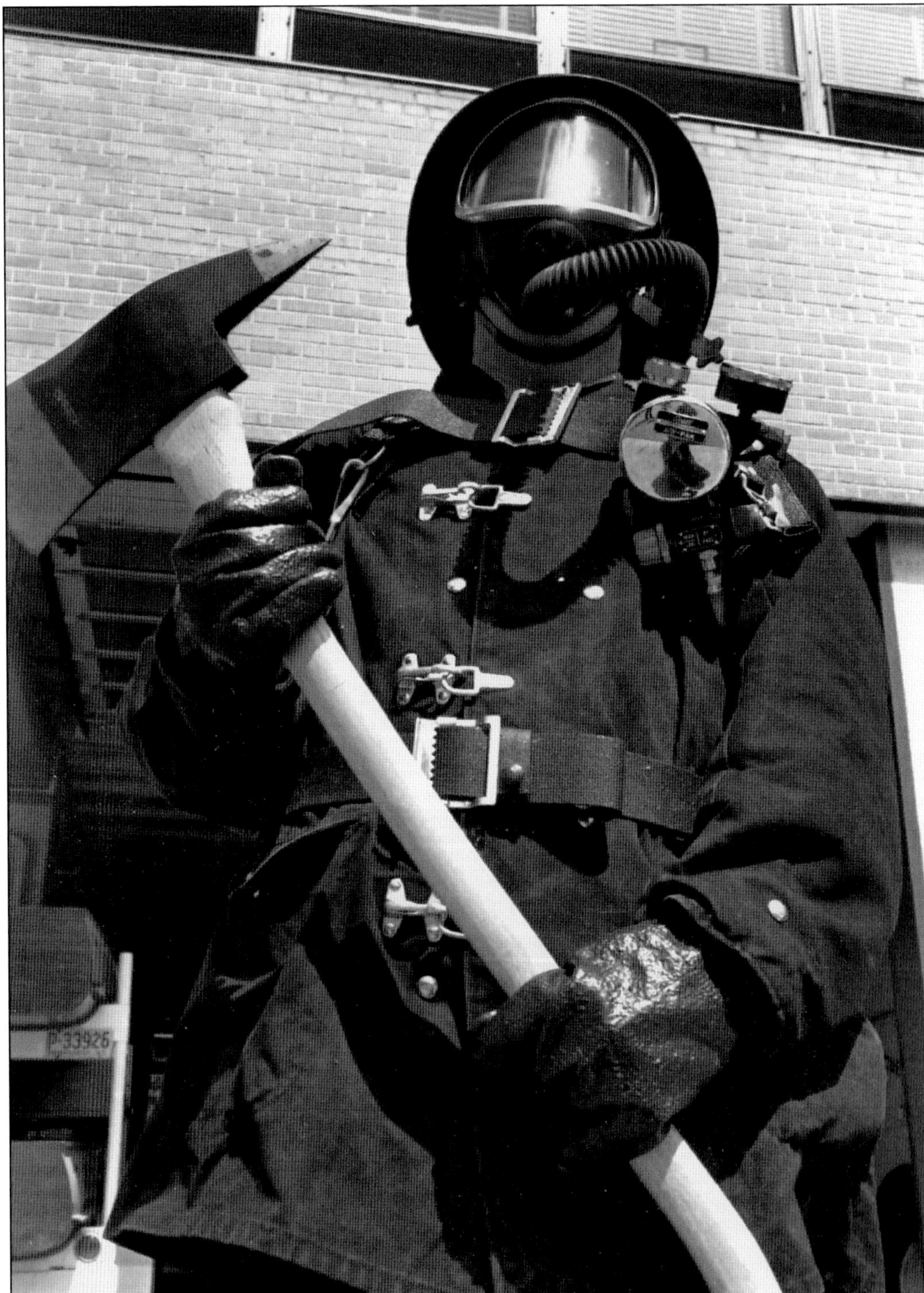

AXE MAN, 1962. Raleigh firefighter E.B. "Buck" King is shown at Station 1 wearing a rubber-lined coat, fireproof gloves, a heavy plastic helmet, and breathing gear. (*RT* photograph.)

IMAGES
of America

RALEIGH *and* WAKE COUNTY FIREFIGHTING
Volume II

Michael J. Legeros

ARCADIA
PUBLISHING

ISBN 978-0-7385-1620-2

Published by Arcadia Publishing
Charleston, South Carolina

Printed in the United States of America

Library of Congress Catalog Card Number: 2003100673

For all general information contact Arcadia Publishing at:
Telephone 843-853-2070
Fax 843-853-0044
E-mail sales@arcadiapublishing.com
For customer service and orders:
Toll-Free 1-888-313-2665

Visit us on the Internet at www.arcadiapublishing.com

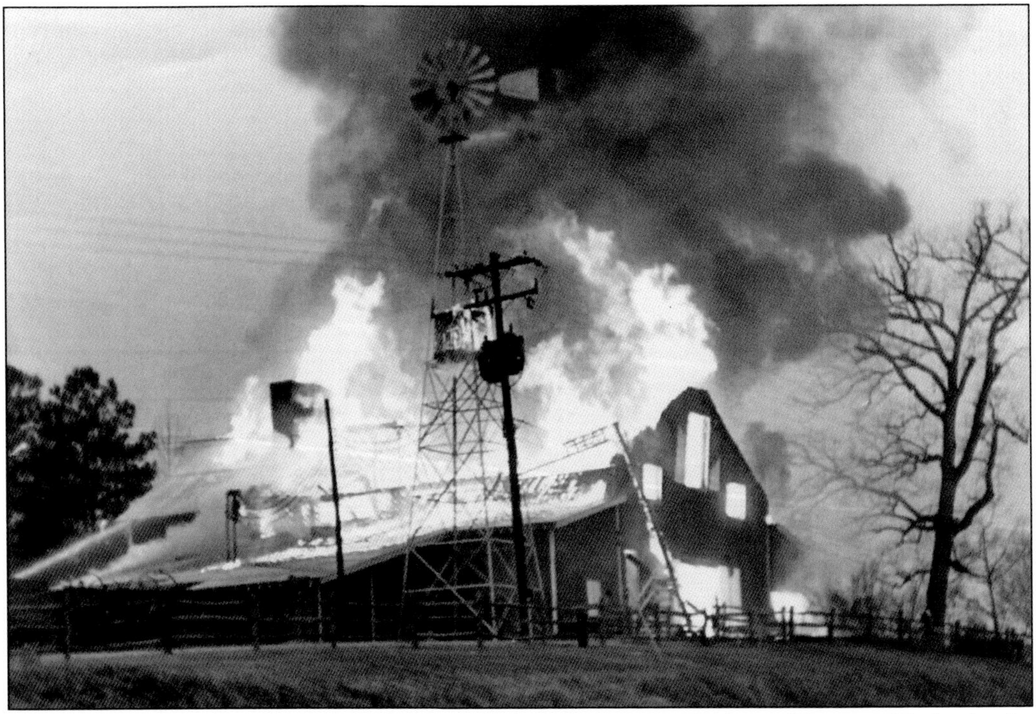

ANGUS BARN, FEBRUARY 7, 1964. Four years after opening on Highway 70 west of Raleigh, the "beef eater's haven" burned to the ground. Constructed to resemble a big red barn, the building had some 9,000 square feet of floor space on each of two levels. A passing truck driver spotted the fire shortly before 7 a.m. Units from Six Forks, Morrisville, Fairgrounds, and Raleigh responded. Water was relayed from nearby Raleigh-Durham Airport. Within an hour, the building was a total loss. The restaurant was rebuilt and reopened on January 27, 1965. (*N&O* photograph by Bob Lynch.)

Contents

ACKNOWLEDGMENTS

The author thanks everyone: Don Adams, Greg Allen and Al Merritt, Hubert Altman, Paul Averette, Thomas Babb, Floyd Bailey, Wade Baker, Barbara and Linwood Barham, Jessica Bennett, Herman Berkhoff, Bob Biggs, Jim Biggs, Bradley Blake, Goley Boggs, Ernest Bridges, Betty Poole Brinkley, Ed Brinson, Courtney Britt, Margaret Broadwell, Peter Brock, Richard Byers, Debbie Byles, Allan Cain, Brooke Cain, Brian Cammack, Gene Capps, Russell Capps, Sue Carr, Clyde Carter, Donnie Carter, Robin Carter, Daryl Cash for his Wake Forest patches, Mike Chambers, Tony Chiotakis, Andy Clark, Judy Cockerham, Michael Cooper, Marty Coward, Jesse Creech, Lloyd Curtis, Don Daniels, Dan Danieley, Mike Davidson, Dolores Dean, Julie Denning, Ross Denson, Mike Dillard, Sanford Dockery, Jeremiah Dodson, Ricky Dorsey, Barry Doyle, J.W. Doyle, Paul Dunwell, Karen Dutton, Jo Edwards, Don Ellington, Georgia Evangelist, Greg Flynn, B. T. Fowler, Earl Fowler, Jo Fulton, Aubrey Gay, Keisha Webb-Gibbs, Robert Gibbs, Dana Goforth, Kat Goodwin, David Grant, Wayne Greer, my sweetie Julie Gresens, Ray Griffiths, Tim Guffey, George Gupton, A.C. Hall, Mark Haraway, Jeff Harkey, Scottie Harris, Dena Height, Carol Baucom Heelan, Jill Highsmith at the Angus Barn, Sue-Lynn Hinson for detective work in Apex, Darrin Holt, John House, Wayne House, E. J. Howard, W. D. Jackson, Dale Johnson for his encyclopedic knowledge of wheeled vehicles, Denise Jones, Jim Jones, Rick Jones, Felix D. Katt, Jimmy Keith, Mike Kemmeries, Leslie Kepley at American LaFrance, Keith King, Gary Knight, Elaine Massengill Kurtz, Michael LeBeau, Julia Ledford and Jason Tomberlin at Hayes Barton Baptist Church, Charles Lloyd and rest of the retired Raleigh firefighter breakfast crew, Keith Longiotti, Freddy Lynn, Steve Massengill and Alan Westmoreland at the North Carolina State Archives, Paul Matthews, Tim Matthews, James Mauldin, Tony Mauldin, Faye McArthur, Scott McCollum, Gary McConkey, Thomas Melville, Gayle Mills, Phil Mitchell, Mom, Dad, Lisa, Tim, Amy Moore, Ann Moore, Harold Moore, Jeff Mullen, Tim Murphy, Elizabeth Reid Murray, Laura New, Jimmy Nobles, Seny Norasingh, Sue Olmsted and the staff at Olivia Raney Library, Cecil Parker, Jan Parker, Robert Pearce, Don Penny, Barrett Penny, George Pleasants, Alton Perry, Chris Perry, Ned Perry, Sidney Perry, Lisa Persinger, Kit Podger, Dewey Poole, David Price, Rodney Privette, Herb Ramsey, Jay Rauer, Johnny Ray, Dr. L. A. Raynor, Bruce Radford, A.C. and Leigh Rich, Tommy Rich, Alexis L. Richardson, David Ritchie, David Roberson, Ron Roof, Chris Rozier, Lisa Sago, Sam's Club photo developing in Cary, my publicist Sara San Angelo, Buddy Scarboro for detective work in Wendell, Mary H. Schaub, Hubert C. Sears, Amy Shekita, Chad Simon, Nick Slobodzian, Shane Snider, La Rue Stalvey, Sue Stell, Ava King Straughn, Lewis Straughn, James and Ruth Swain, my editor Maggie Tiller, Jimmy Thompson, Hermann Trojanowski, Dwayne Vaughn, Tom Vaughn, Mike Wallace, Rodney Warner, Jimmy Warren, Dusty Wescott, Kaye Whaley, Greg Wiggs, Chris Wilson, Daphne Wilson, Lee Wilson for contribution *and* distribution, Jim Wise, Steve Woodall, Frank Woods, Tom Zumbado , and every other firefighter, photographer, editor, archivist, librarian, historian, buff, fan, friend, and family member whose support and contributions were absolutely invaluable in creating both volumes of *Raleigh and Wake County Firefighting*.

Introduction

This second collection of historic firefighting images concentrates on the years 1940 to 1990, five formative decades for fire protection in North Carolina's capital county. During this time, the Raleigh Fire Department quadrupled in size, smaller municipal departments modernized, and rural departments organized. Raleigh's fire department was firmly established by 1940 with five stations and 56 full-time firemen operating five engines and two truck companies. They were 27 years old as a "paid department" and decades past their horse-drawn days. Outside of Raleigh, many of Wake County's smaller towns had purchased fire equipment by 1940. These ranged from hand-drawn hose reels to early-model pumpers. They had fire chiefs with slim salaries and volunteers to respond from their homes and businesses. (For their service, the firemen received a couple of dollars per call or exemptions on certain taxes.) By the late 1940s and early 1950s, these towns—Apex, Cary, Fuquay Springs, Wake Forest, Wendell, and Zebulon—each had a brand-new, fully equipped fire engine with a full complement of volunteer firefighters. Wake Forest even had a second fire company, organized in 1942 to protect African-American residents.

Beginning in the 1950s, formal fire protection became available outside of Raleigh and those small-town centers. A rural fire department program enabled any group of citizens to organize a volunteer department. The county Office of Civil Defense offered funding, equipment, and a countywide radio network. Lower insurance rates were an added incentive. And so they formed—inside, outside, and entirely removed from Wake's municipalities. Residents of Garner, Knightdale, Morrisville, and Rolesville formed their town's first fire companies in the 1950s. In the suburbs of Raleigh, the New Hope, Six Forks Road, and Western Boulevard fire departments formed in the late 1950s; the Durham Highway and Fairgrounds fire departments forming in the early 1960s. Who were these volunteers? Six Forks Road members in 1958 included a grocer, barber, detective, telephone man, lawyer, schoolteacher, florist, mail carrier, heating contractor, two mechanics, a Civil Defense official, and several state workers.

In those towns that already had fire departments, rural service was started for farmers and others located outside the incorporated limits of Apex, Fuquay Springs, Wake Forest, Wendell, and Zebulon. Though separate entities with separate equipment, most were staffed by the same set of firemen. In Cary, the volunteer fire department split in 1961 into separate groups with separate equipment and personnel. They became the Cary and Yrac fire departments. Other communities that formed fire companies were Stony Hill in 1958, Swift Creek in 1960, Bay Leaf in 1961, and Fairview in 1963. The trend continued into later decades, as well, with the formation of the Falls, Holly Springs, and Hopkins Fire Departments and the opening of "substations" or second stations in places like Carpenter, Panther Branch, and Wake Crossroads.

Once a rural fire department was formed, volunteers and community members built the necessary infrastructure. They held fund-raisers and gave money, elected directors, and established by-laws. They donated barns and garages for storing trucks and built tin sheds and block buildings as the first firehouses. They even constructed their own fire apparatus, typically converting surplus vehicles into home-built pumpers and tankers. The Fairview Fire Department was a typical example. Their first fire engine was built from an ex-weapons carrier. Members added a water tank and pump, mounted two taillights as warning lights, and applied a coat of donated paint in a chicken shack. It was stored in a building whose doors were untied, laid flat, and driven over when they responded to a call.

This book is dedicated to those volunteer firefighters, both past and present, from the first hand-engine crews in Raleigh to the hundreds of men and women still serving Wake County today. They rarely receive the recognition they're due.

Please note the following abbreviations in credits: AFD—Courtesy of Apex Fire Department; BLFD—Courtesy of Bay Leaf Fire Department; CFD—Courtesy of Cary Fire Department; DAH—Courtesy of North Carolina State Archives; EWFD—Courtesy of Eastern Wake Fire Rescue; FFD—Courtesy of Falls Fire Department; FVFD—Courtesy of Fuquay-Varina Fire Department; *FVI—Fuquay-Varina Independent*; GFD—Courtesy of Garner Fire Department; *GLF—Gold Leaf Farmer*; HFD—Courtesy of Hopkins Fire Department; MRFC—Courtesy of Morrisville Rural Fire Company; *N&O—News & Observer*; RFD—Courtesy of Raleigh Fire Department; RRFD—Courtesy of Rolesville Fire Department; *RT—Raleigh Times*; SCFD—Courtesy of Swift Creek Fire Department; WFD—Courtesy of Wendell Fire Department; *WW—Wake Weekly*; WWFD—Courtesy of Western Wake Fire Rescue; ZFD—Courtesy of Zebulon Fire Department.

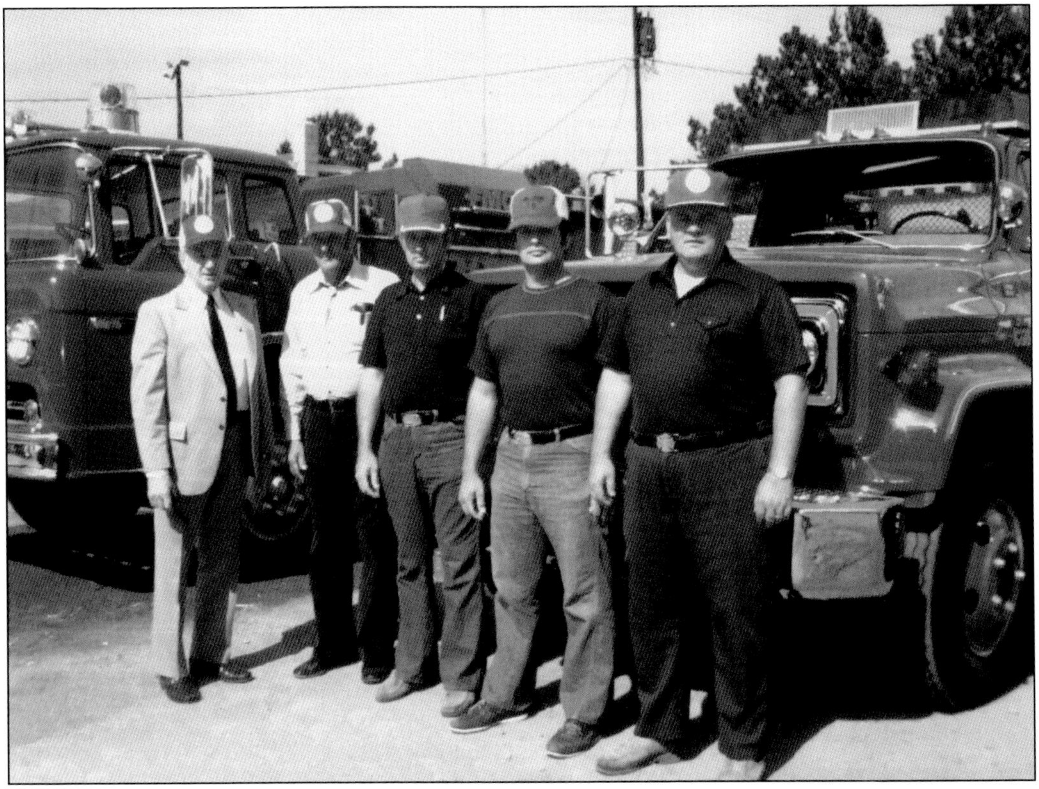

STATE FAIR FIRE BRIGADE, OCTOBER 1979. Pictured, from left to right, are fair manager Art Fizer and brigade members Earl Coley, Donald Sykes, Paul Averette, and James Warren. Formed in the 1960s, the brigade provides fire protection and conducted safety checks, and later, emergency medical services. Their first fire truck was borrowed from Garner. Later apparatus was donated by dealers, notably Slagle's Fire Equipment in South Boston, Virginia. (RFD.)

One

FIRE!

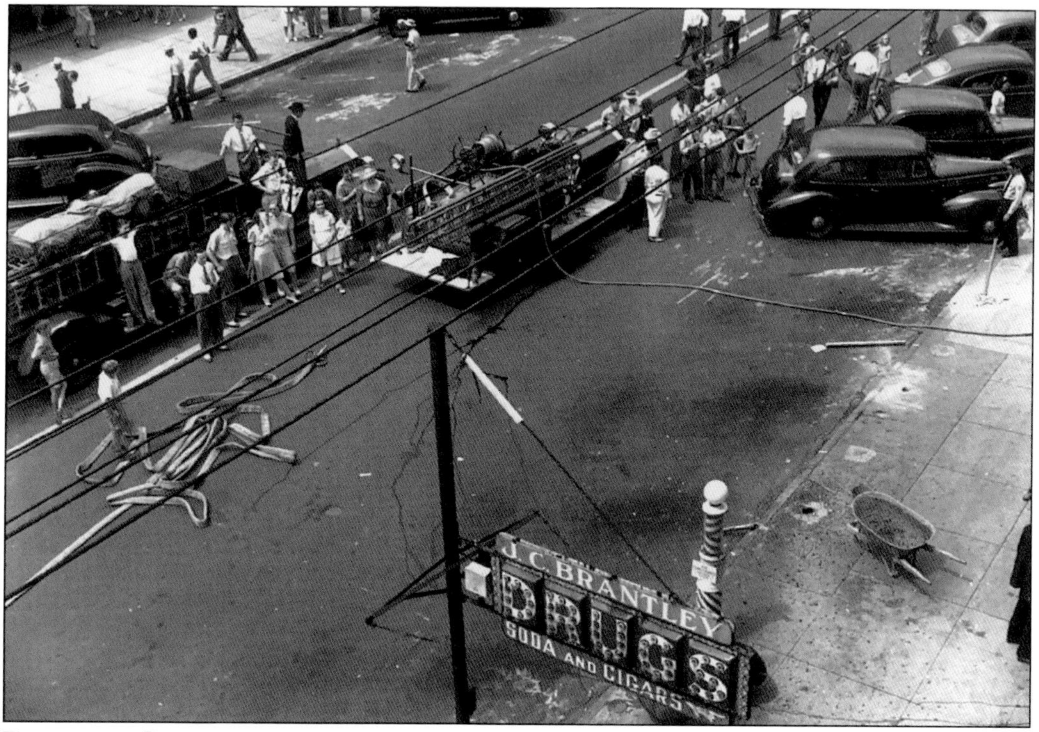

DOWNTOWN RALEIGH, AUGUST 24, 1941. Spectators stand beside fire hose splayed along the 100 block of Fayetteville Street. (*N&O* photograph.)

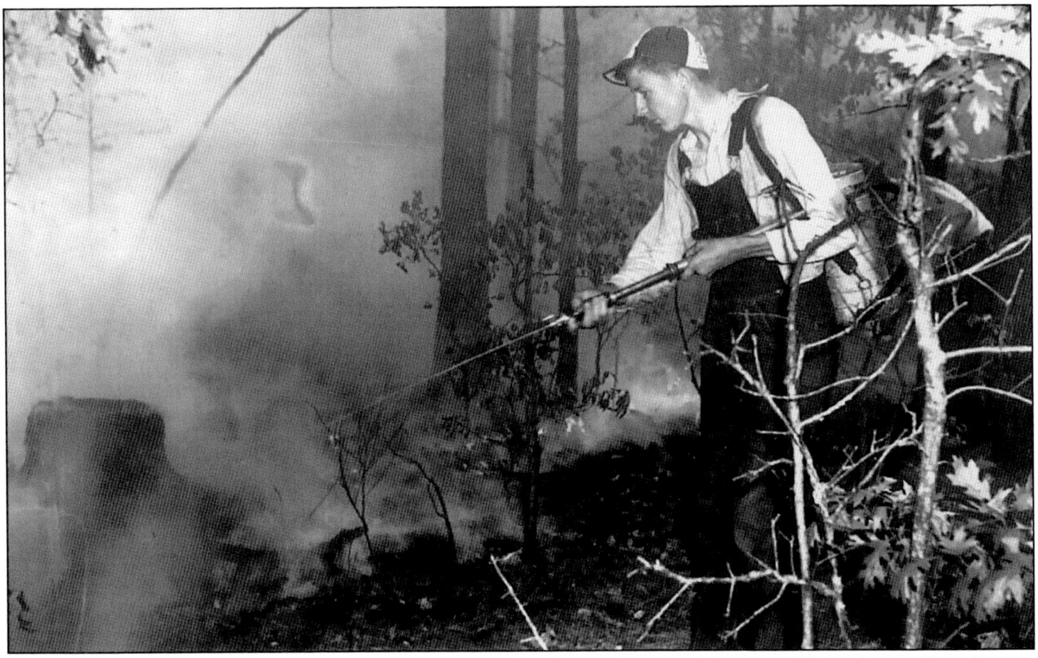

Woods Fire, April 22, 1941. Willis Wester battles a 50-acre blaze on Rhamkatte Road near Macedonia. Supervised by Wake Fire Warden R.L. Lassiter, residents had the fire nearly under control by the time the Civilian Conservation Corps arrived. (*N&O* photograph.)

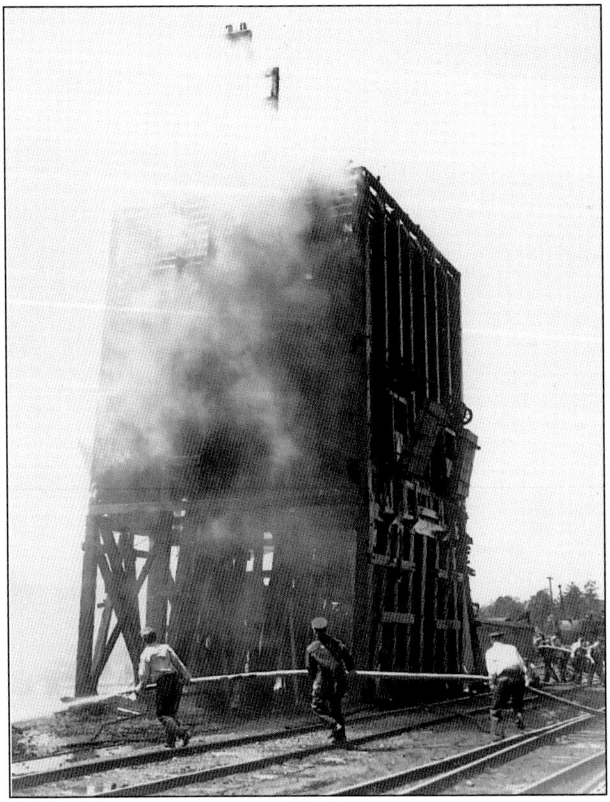

Coal Bin, May 9, 1941. This stubborn fire at the Norfolk-Southern Railroad yards brought discussions of dynamite as a solution. Using all available equipment from Station 1, firefighters poured water for six straight hours. The fire was eventually controlled without the use of explosives. Damage was estimated at $10,000. The railroad yards were located in the northern section of the city, near Roanoke Park. (*RT* photograph.)

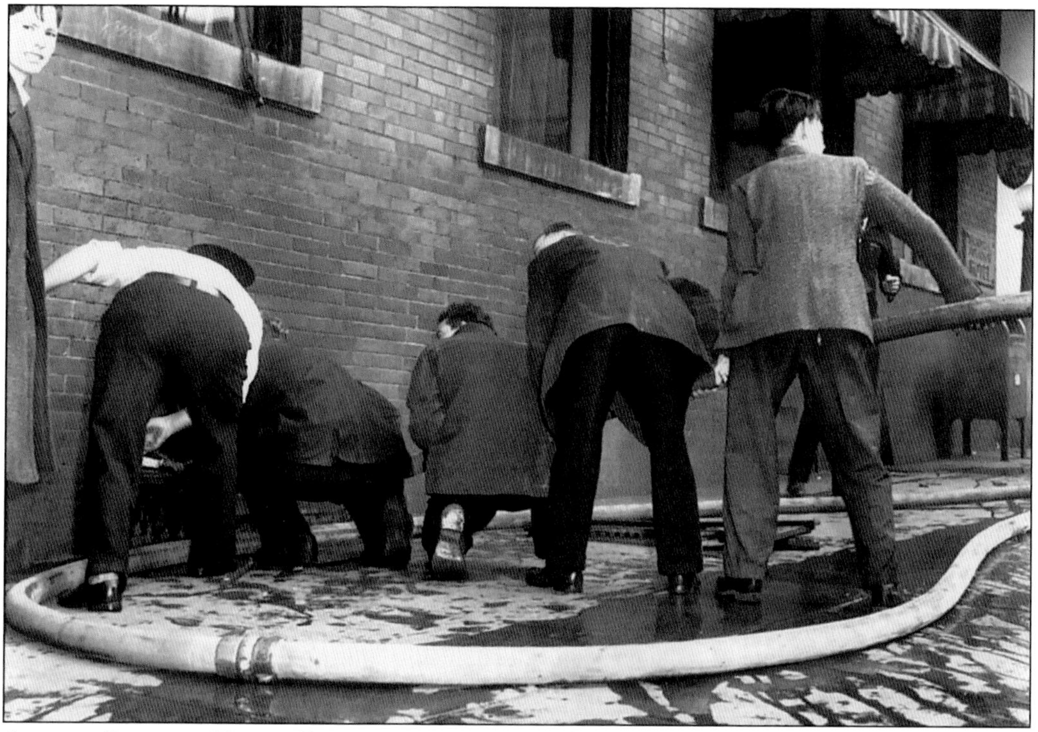

Andrew Johnson Hotel, February 28, 1942. Raleigh firefighters battle a basement blaze at 100 W. Martin Street that started in a storeroom but spread quickly to the barbershop, workshop, and pressing room. Though smoke poured throughout the hotel, flames never reached the first floor. Four hose lines were used, along with gas masks and a new "portable light transmitter." (*N&O* photograph by Robert Johns.)

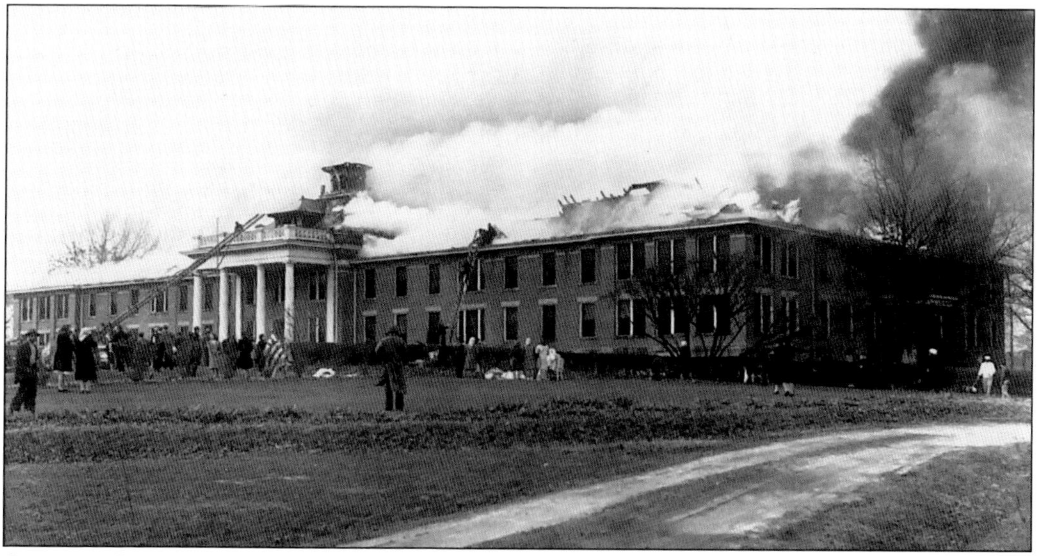

County Home, July 6, 1943. Flames sweep the roof of the Wake County Home for the Aged and Infirm on E. Whitaker Mill Road in Raleigh. The fire was discovered by a cook who rang the dinner bell as an alarm. Civil Defense workers, neighbors, and other volunteers helped remove the 134 residents. (*N&O* photograph by Lawrence Wofford.)

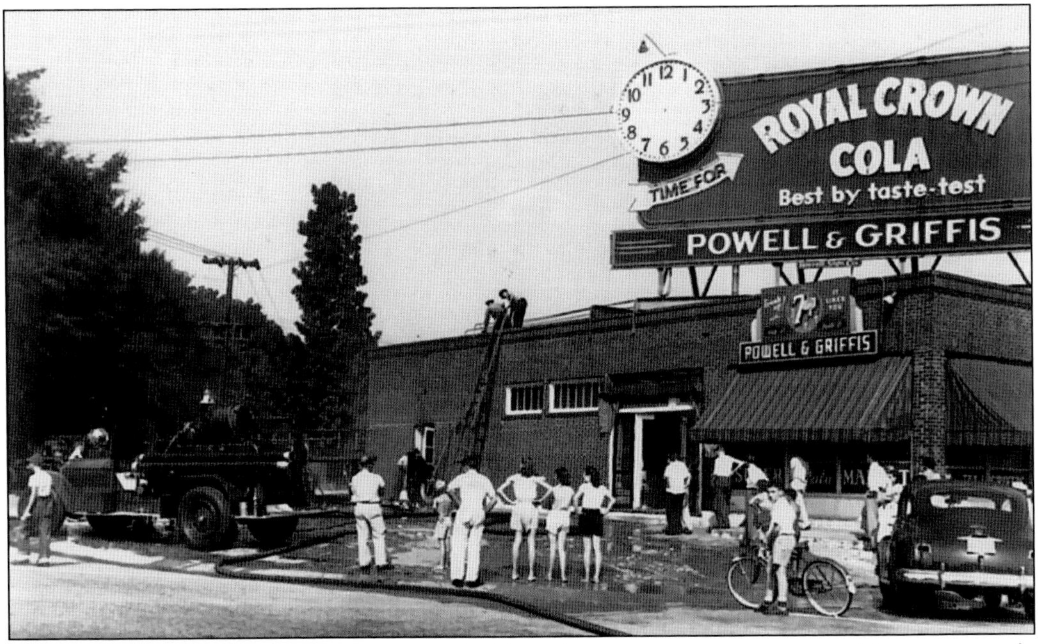

Five Points Market, August 10, 1949. Though the store proper escaped harm, an early afternoon fire caused considerable damage to the storage room at Powell and Griffis Market in Raleigh. Two pumpers and a ladder truck were dispatched to the 3:14 p.m. call at 1700 Glenwood Avenue. (*N&O* photograph.)

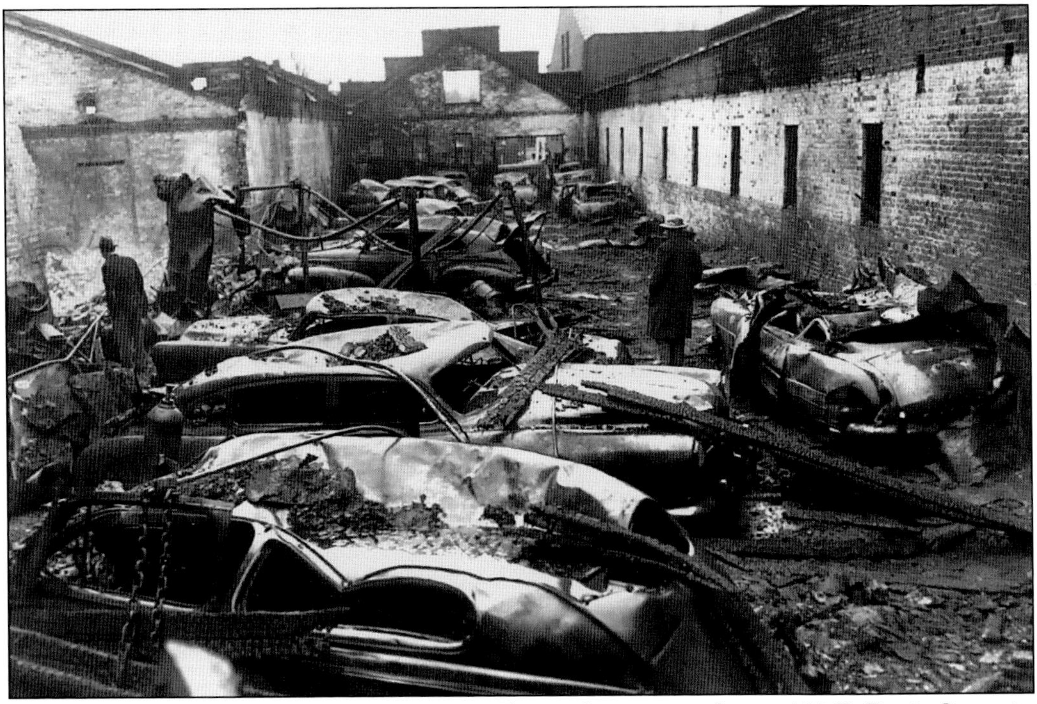

K&W Motor Company, March 3, 1952. This early-morning fire at 118 E. Davie Street in Raleigh destroyed 29 automobiles, new and used. Seven companies battled the 2:30 a.m. blaze in a bitter cold rain. (*N&O* photograph by Lawrence Wofford.)

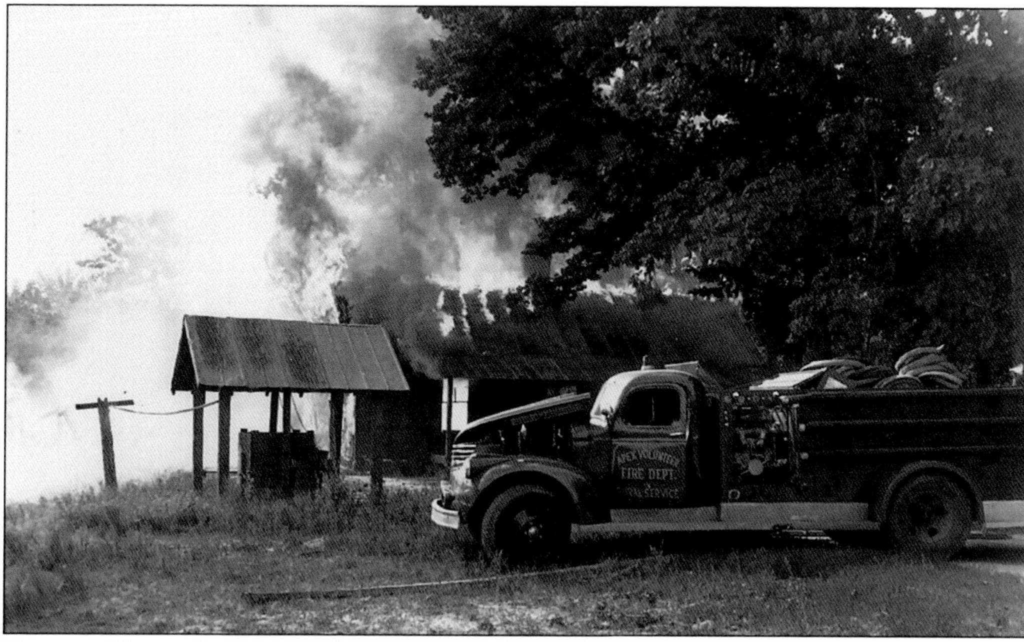

Tenant Home, September 6, 1954. The Apex Fire Department's rural truck is pictured pumping on the farm of John H. Luther on Route 2. The house was more than 100 years old and the occupants saved only the clothes they were wearing. (*N&O* photograph by Lawrence Wofford.)

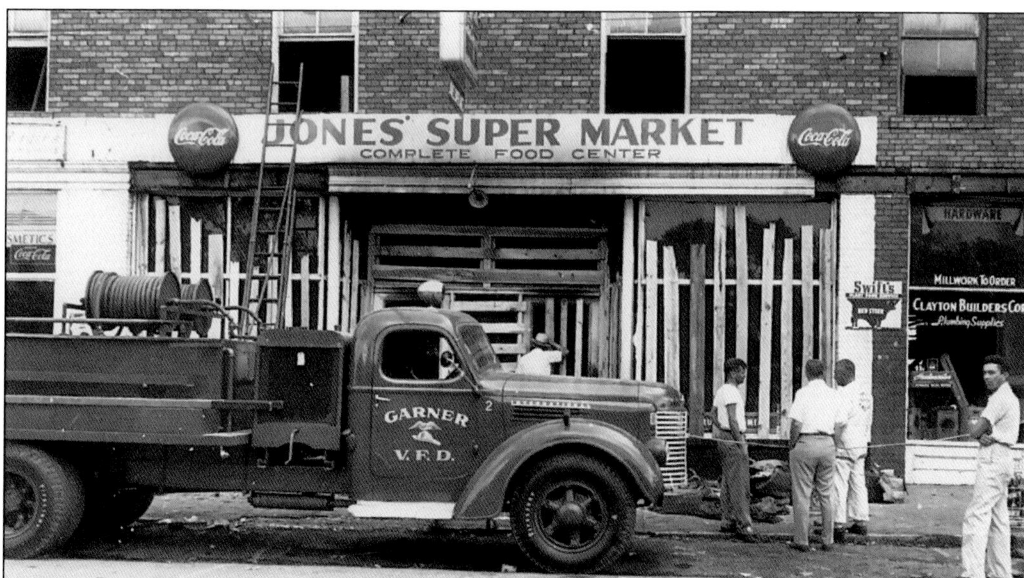

Jones Super Market, July 24, 1955. Firefighters from both Raleigh and Cary assisted the Garner Volunteer Fire Department as a 4 a.m. blaze threatened the entire business district. Townspeople helped, too, forming a bucket brigade to fill water tanks. After the flames were finally extinguished, the wives of the men fighting the fire served them a breakfast at the nearby American Legion headquarters. Damage to the interior of the store was estimated at $20,000. (*N&O* photograph.)

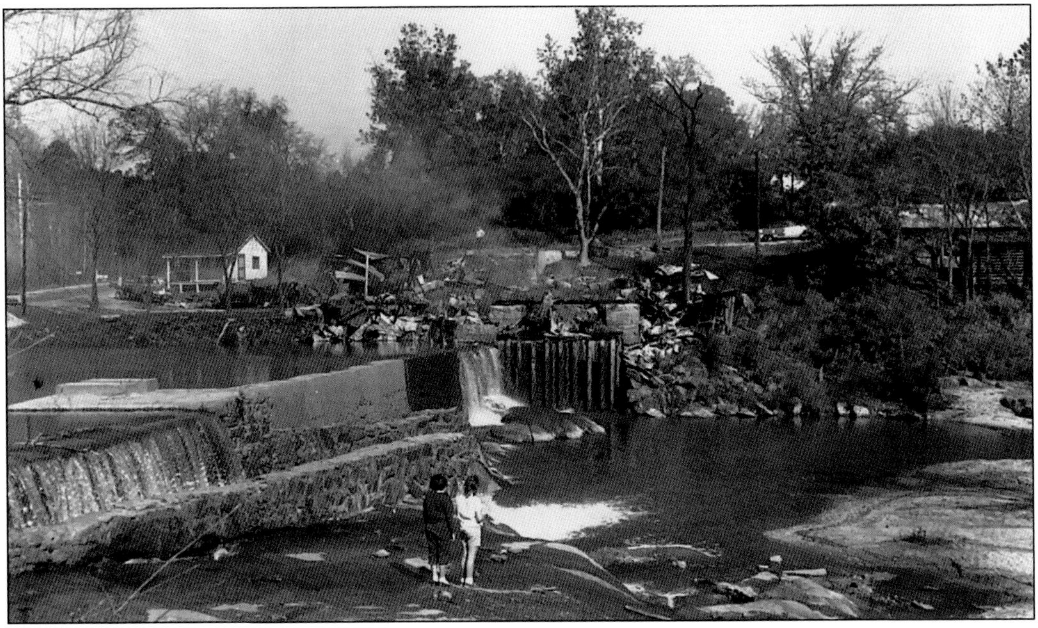

LASSITER'S MILL, NOVEMBER 8, 1958. Only the waterfall and a pile of debris remain at one of Wake County's oldest businesses. Flames were seen breaking through the roof of the mill around 2 a.m. About 25 firefighters from Six Forks and New Hope were unable to save the structure but did prevent a nearby lumber mill from burning. Located on Crabtree Creek, the site was also a popular picnic ground and swimming hole. The mill was founded in 1907 by Cornelius J. Lassiter and supplied meal to most grocery stories in Raleigh. (*N&O* photograph by Ed Chabot.)

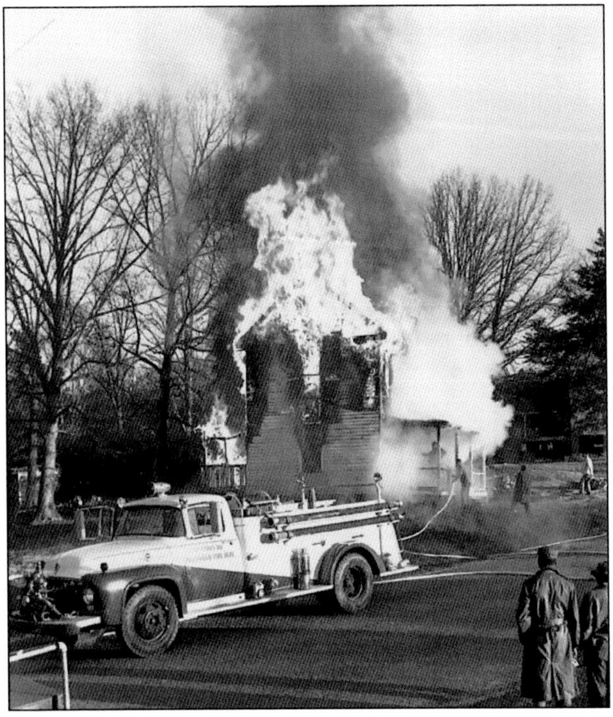

LASSITER'S HOME, FEBRUARY 12, 1957. One year earlier, this four-room frame home owned by the milling company burned to the ground. A family of five escaped through a second-story window after a fire pot exploded. The husband, who was thawing pipes, suffered burns to his ankles. The rest of the family was uninjured. Located up the hill from the mill, the house was built in 1906 and had been owned by the milling company since 1948. It was one of the first major fires for the pictured Six Forks Road Fire Department. (*RT* photograph by Ed Chabot.)

EDENTON STREET METHODIST CHURCH, JULY 28, 1958. By the time of this photograph, flames "had toppled the 200-foot steeple of the historic church at 228 W. Edenton Street in Raleigh. The flaming spire collapsed at 8:10 p.m., about an hour after being struck by lightning. The blaze drew thousands of spectators. Off-duty firemen were called to the scene. The Cary and Garner Fire Departments also assisted. Built in 1811, the church was the first place of worship in the city. A fire in 1839 destroyed that building. The church was rebuilt in 1841 and again in 1887. A separate Sunday school building, built in 1838, received water damage; the sanctuary, renovated in 1951, was destroyed. The loss was estimated at a minimum of $500,000. (*N&O* photographs.)

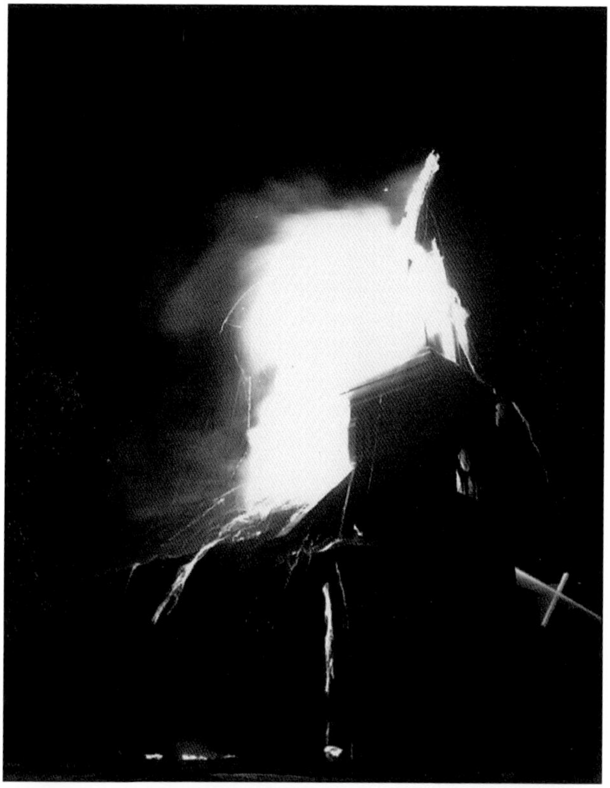

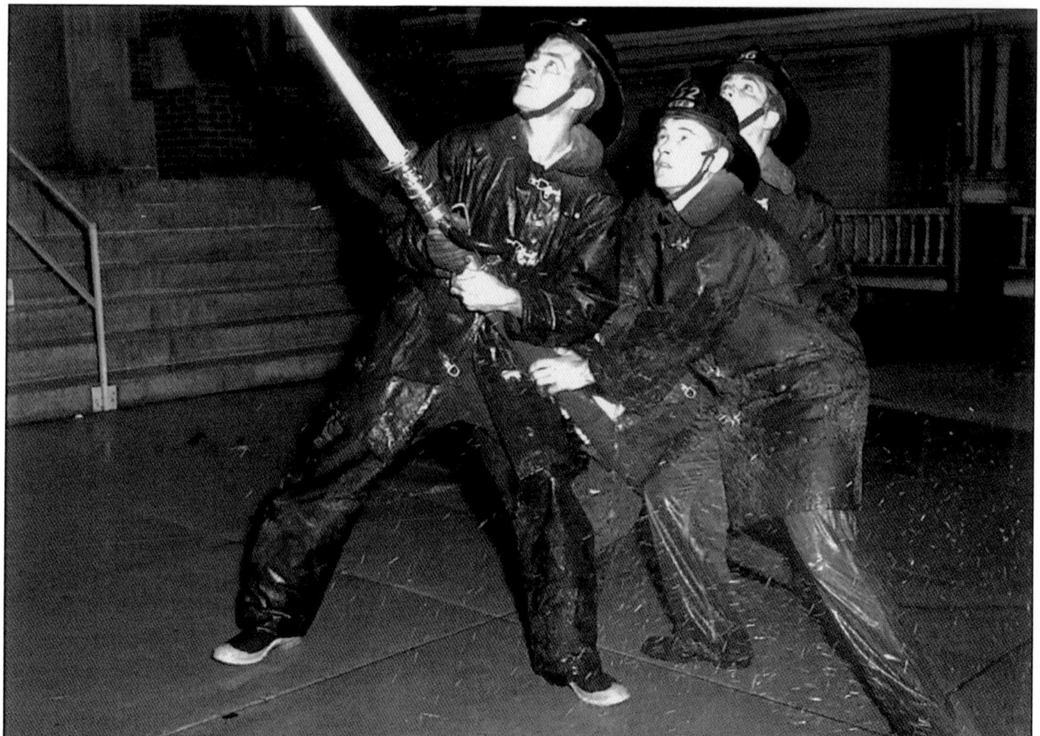

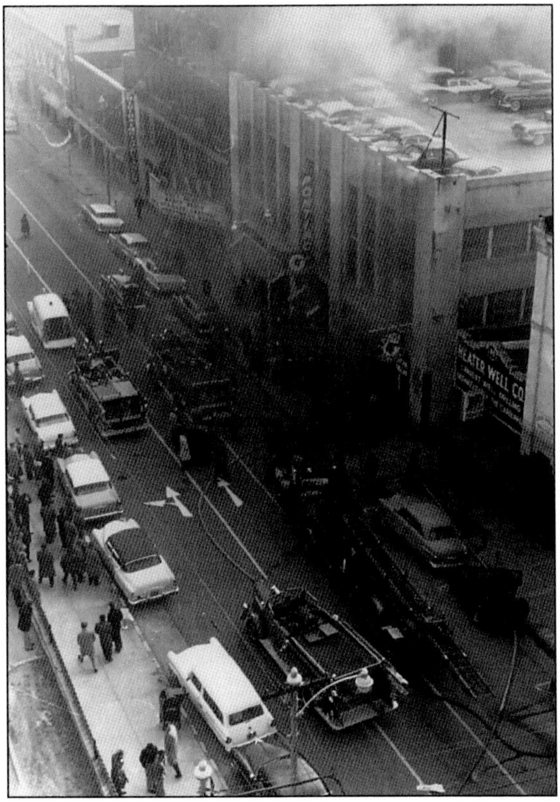

CONN-GOWER PONTIAC, FEBRUARY 18, 1958. Gasoline that had been splashed onto a heater started this afternoon fire at 310 S. Salisbury Street in Raleigh. The resulting explosion and fire damaged an automobile, desk, and tool rack. Despite freezing winds, some 300 spectators gathered on the sidewalk to watch the 4:15 p.m. blaze. Four units from nearby Station 1 responded and had flames extinguished within 20 minutes. Though a mechanic was slightly burned, damage to the structure was minimal. Owner Charles Conn described his building as "fire resistant to the last degree." (*N&O* photograph.)

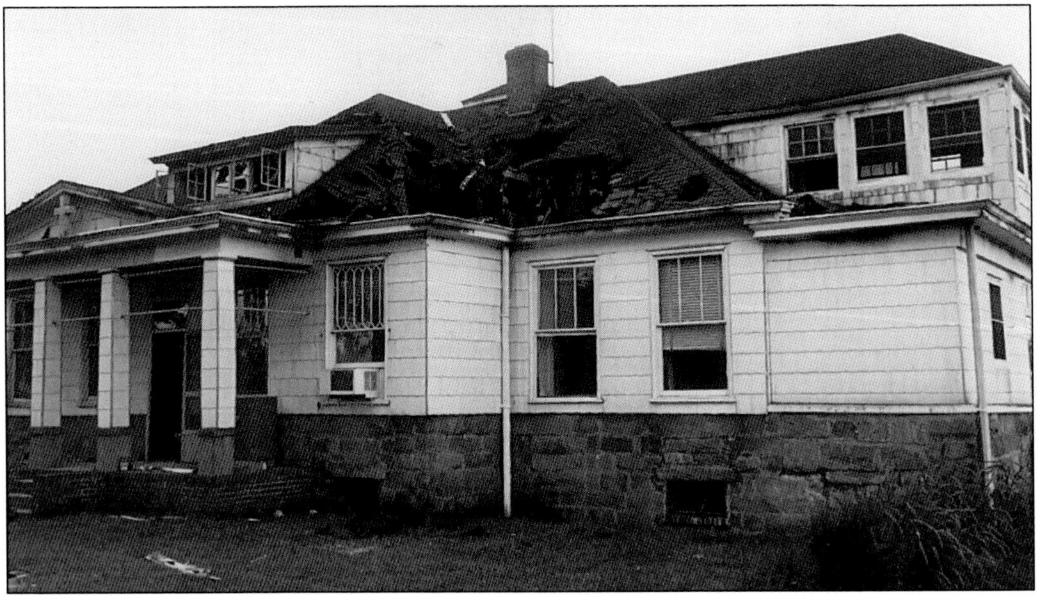

"WASPS GONE; BUILDING, TOO," JULY 13, 1961. That was the newspaper headline after Rev. Raymond J. Donohue burned 14 wasp nests at the Catholic Orphanage in west Raleigh. He also destroyed his office and living quarters. The 20-room, two-story building also housed the Nazareth Post Office. Firefighters laid some 5,400 feet of hose to the nearest hydrant on Western Boulevard. (*N&O* photograph.)

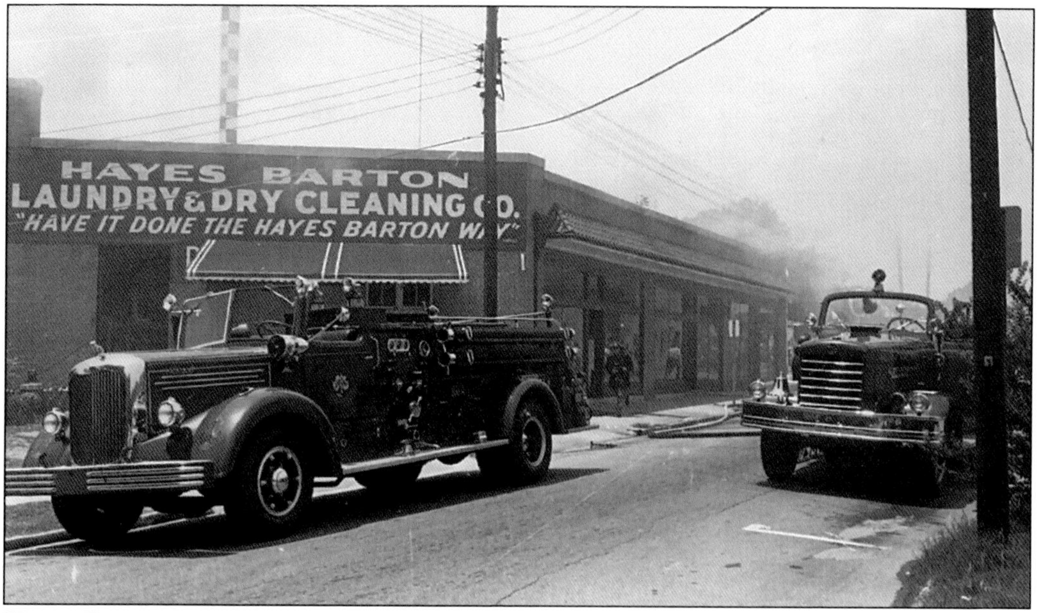

HAYES BARTON DRY CLEANERS, JULY 23, 1962. Confined to the ceiling and attic of 1917 Fairview Road in Raleigh, this fire started shortly before noon at the Five Points laundry. Neither any clothing nor cleaning equipment was damaged. Engine 6 (left) and Engine 4 are pictured, a 1950 Mack and 1957 FWD, respectively. (*N&O* photograph by Ed Chabot.)

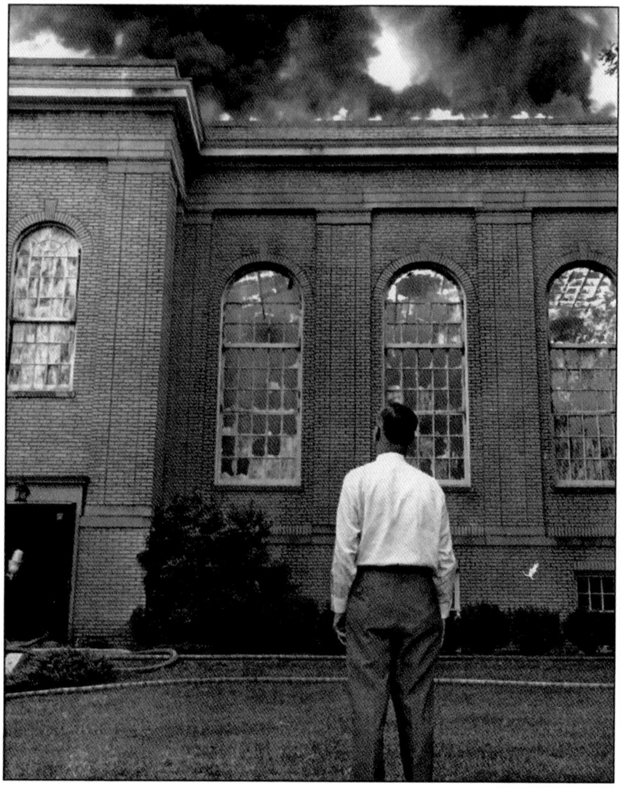

HAYES BARTON BAPTIST CHURCH, SEPTEMBER 5, 1962. Rev. John Kincheloe watches as his 35-year-old church burns at 1800 Glenwood Avenue in Raleigh. Discovered by a member of the congregation as they drove through the Five Points intersection, the 6 a.m. fire raged through the sanctuary for over an hour. Firefighters concentrated on preventing the flames from spreading to the relatively new educational wing. Three were injured after being struck by parts of the falling roof; George Coats, Leland Frazier, and Reginald Poole were treated and released at Rex Hospital. They also got their pictures in the paper. (*RT* photograph by Ken Cooke.)

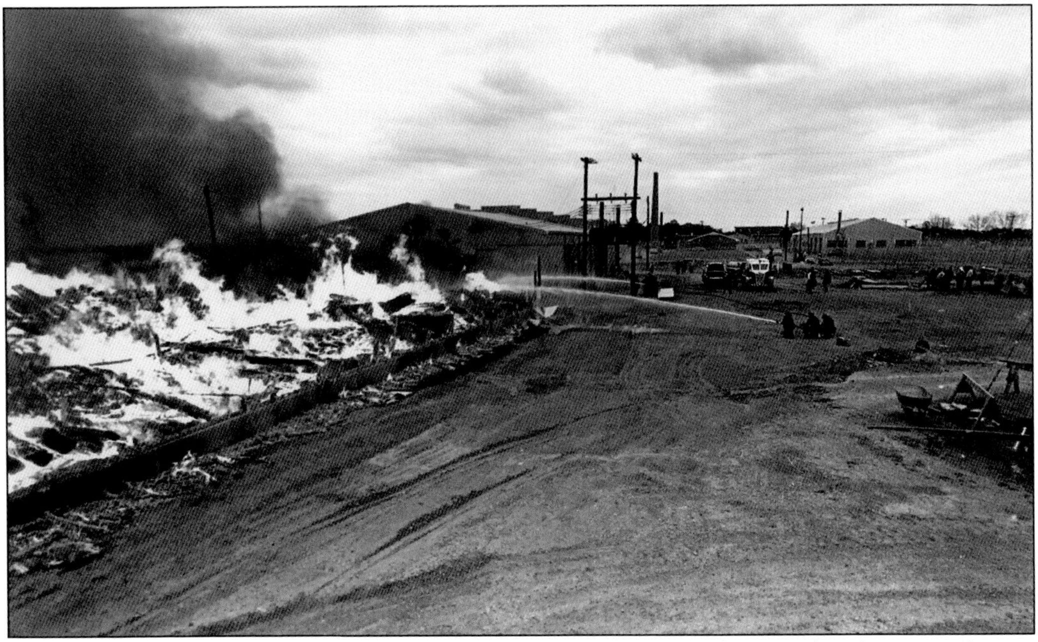

Camp Polk, November 14, 1963. Discovered about 3 p.m. in the woodworking warehouse, these wind-fanned flames soon spread to the mattress plant. Both buildings on Blue Ridge Road were destroyed. The Fairgrounds Fire Department responded along with two Raleigh units. Originally a tank training base during World War I, Camp Polk was converted to a prison work farm in the 1920s and became a youth detention facility in the 1960s. (*N&O* photograph.)

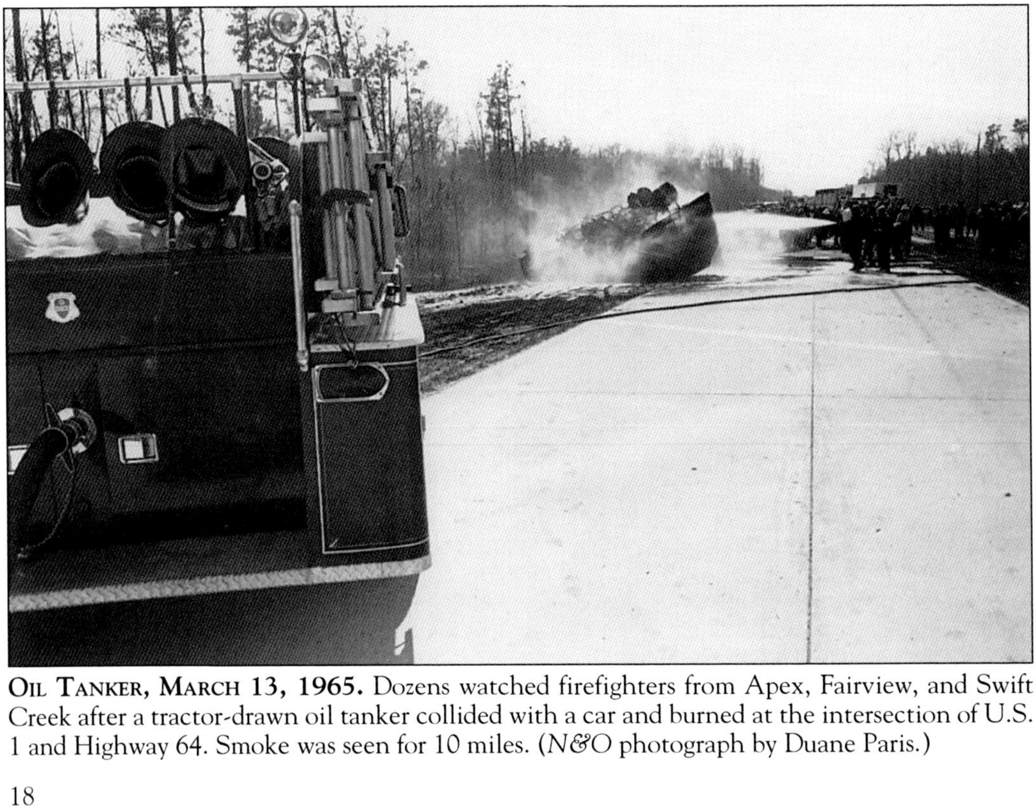

Oil Tanker, March 13, 1965. Dozens watched firefighters from Apex, Fairview, and Swift Creek after a tractor-drawn oil tanker collided with a car and burned at the intersection of U.S. 1 and Highway 64. Smoke was seen for 10 miles. (*N&O* photograph by Duane Paris.)

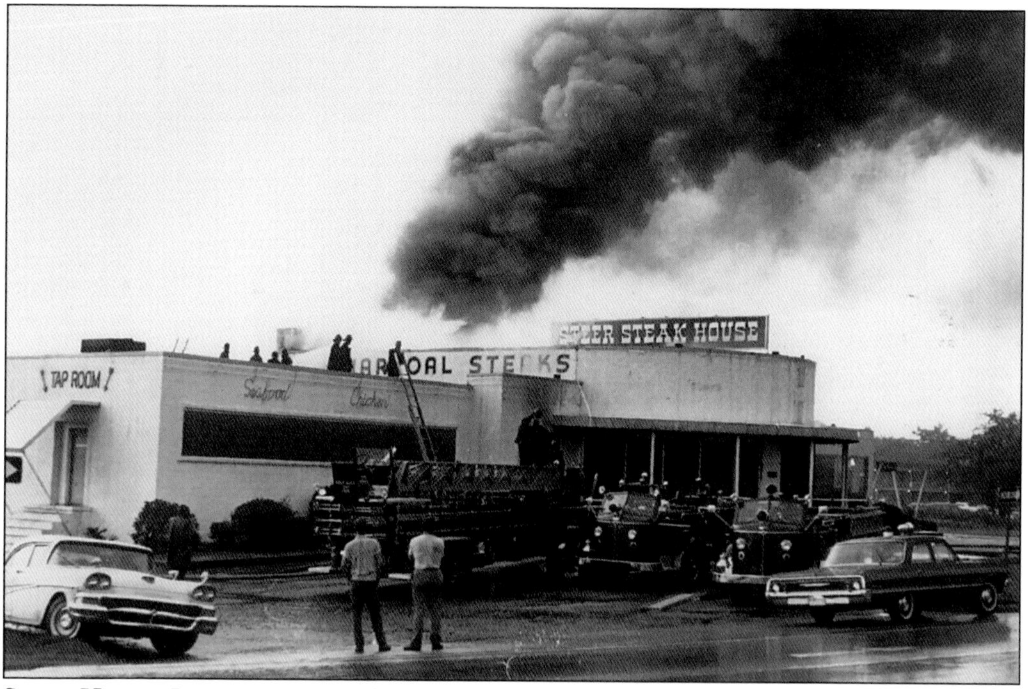

Steak House, July 28, 1965. This fire at the Steer Steak House at 1625 North Boulevard in Raleigh was first spotted by a passing motorist. Three pumpers, two ladder trucks, a foam truck, and a rescue unit responded. It was reported at 3:30 a.m. (*N&O* photograph by Lawrence Wofford.)

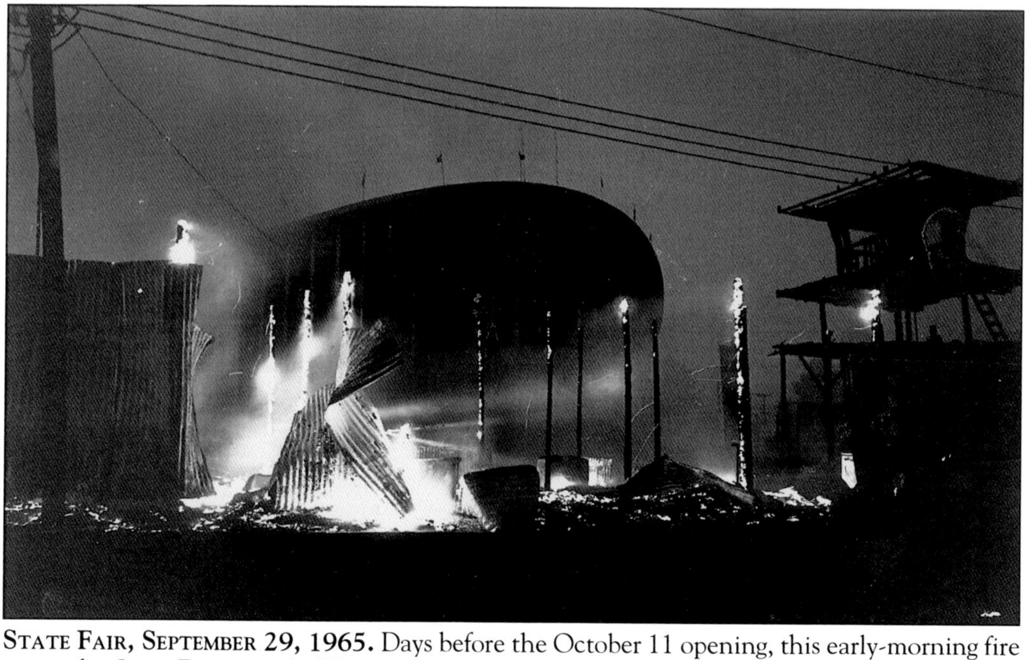

State Fair, September 29, 1965. Days before the October 11 opening, this early-morning fire swept the State Fairgrounds. Eleven concession stands and a public address system tower were destroyed. Both Fairgrounds and Raleigh responded to the 4 a.m. alarm. Dorton Arena is visible in the background. (*N&O* photograph by Lawrence Wofford.)

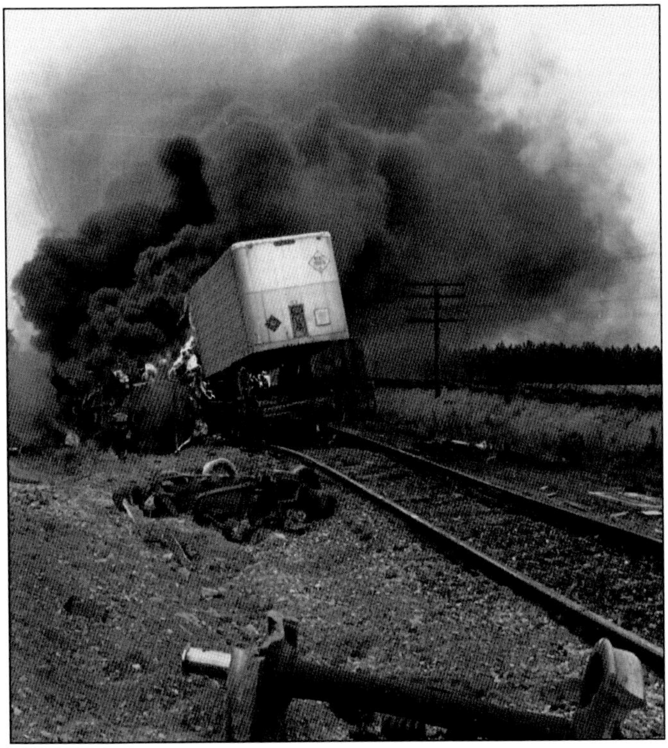

Derailment, April 23, 1966. Flames shot 50 feet into the air after eight cars of a Seaboard Air Line Railroad freight train derailed and burned just south of Wake Forest. Firefighters from Wake Forest, New Hope, Rolesville, Durham Highway, and Raleigh battled the Saturday afternoon blaze as spectators filled a nearby embankment. Some brought picnic suppers. One car carrying frozen poultry was later condemned and every fireman took home a case of frozen chicken. (*N&O* photograph by Martin Rogers.)

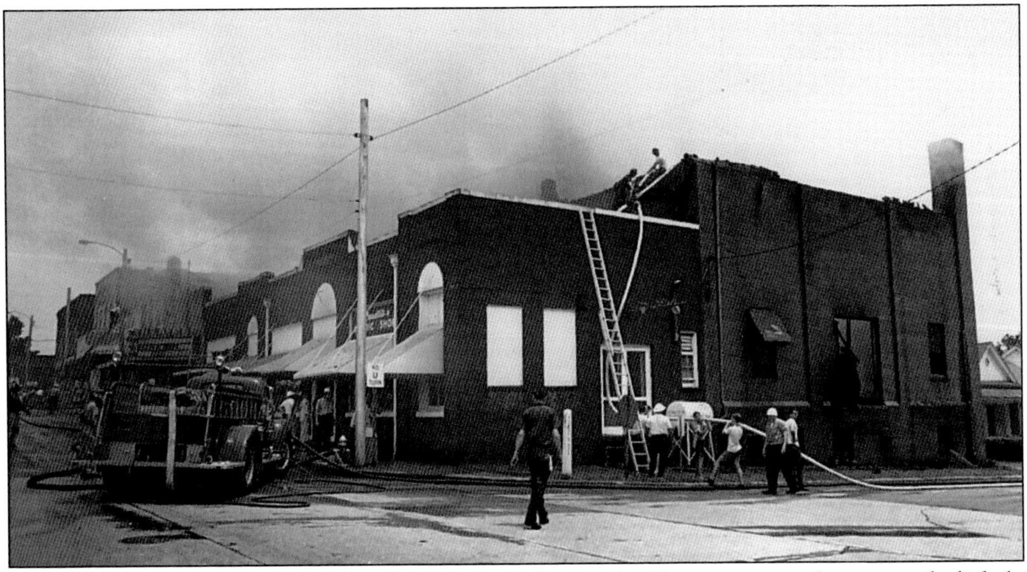

Forest Theater, July 1, 1966. Fire gutted this Wake Forest movie theater and slightly damaged three smaller downtown shops. The 10:15 a.m. blaze, battled by seven departments from Wake and Franklin counties, was brought under control by 11:20 a.m. Bystanders also assisted, including a merchant who grabbed a fire hose to spray water on the roof. Recorder's Court clerk Viola Wilson, who took the alarm, told a newspaper "It's the biggest thing that's happened at Wake Forest since Sherman came through." (*N&O* photograph by Martin Rogers.)

PARK CENTRAL HOTEL, MAY 26, 1967. A burning mattress filled fourth-floor halls with smoke at 138 W. Martin Street in Raleigh. Damaged to the historic hotel was confined to a single room. Completed in 1893 as the Park Hotel, the seven-story structure at the corner of Martin and McDowell Streets was designed by A.G. Bauer, who also provided the working drawings for the Governor's Mansion. The hotel was demolished in 1975 to make way for a parking lot. (*RT* photograph.)

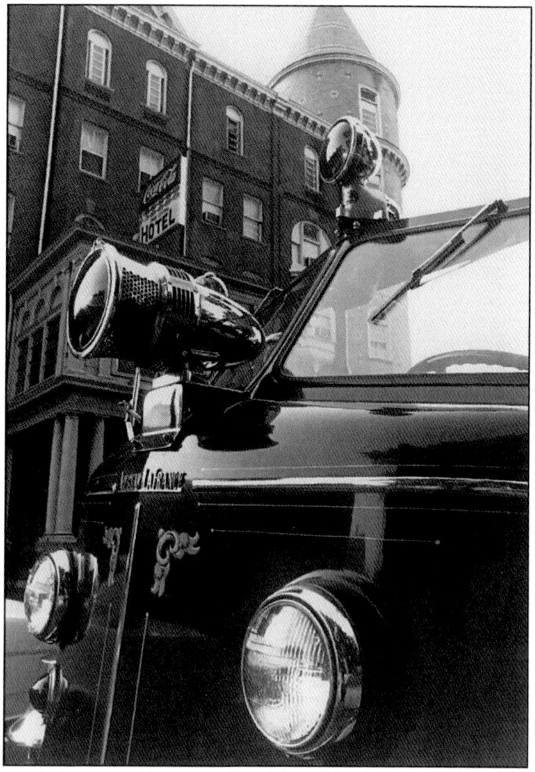

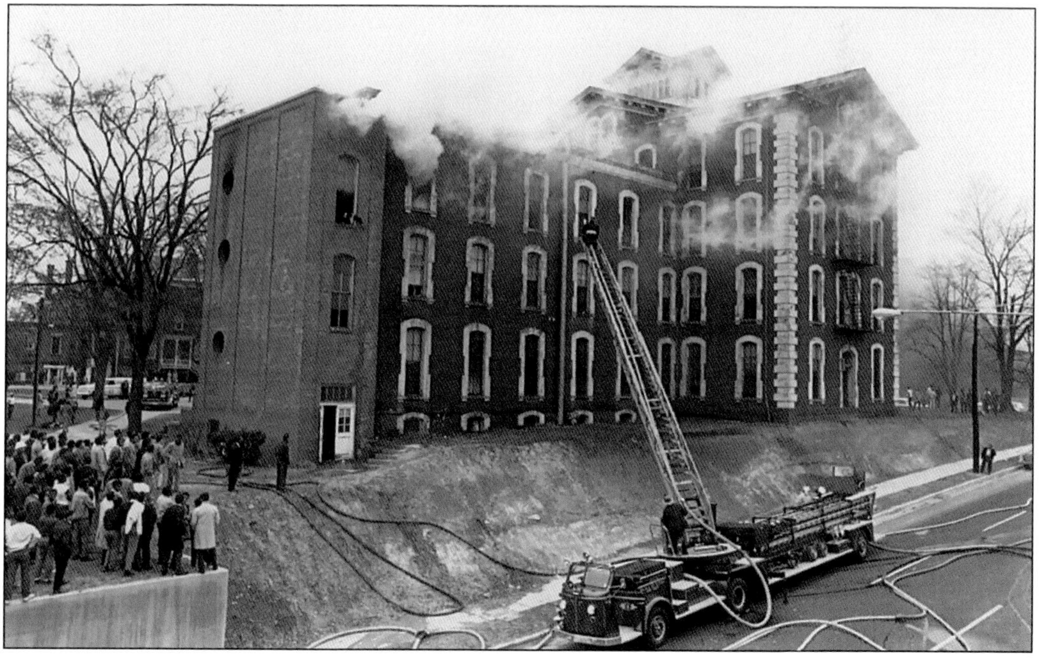

ESTY HALL, DECEMBER 13, 1968. This 97-year-old Shaw University dormitory sustained damage to two rooms and a hallway. Five units responded to the 3:27 p.m. alarm. The third-floor fire was brought under control within 15 minutes. (*N&O* photograph by Ken Cooke.)

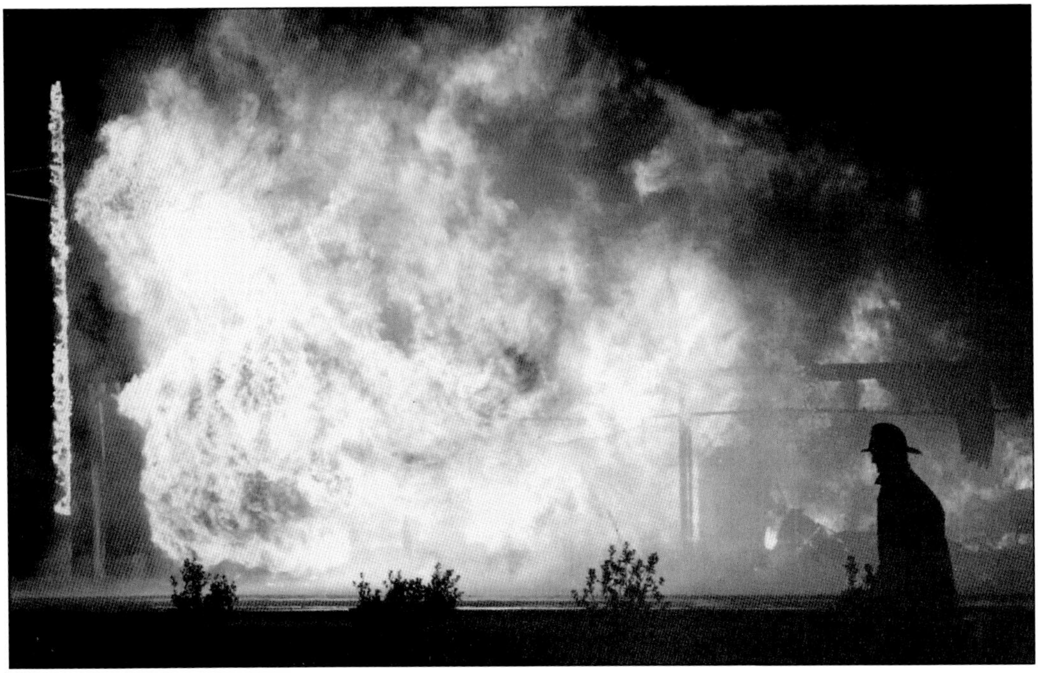

Stephens Supply Company, December 14, 1968. The hardware department of the Fuquay-Varina business was destroyed by this Saturday-night blaze. Reported at 11:30 p.m., the town's fire departments received assistance from Raleigh, Apex, Fairview, Swift Creek, and Angier-Black River in Harnett County. (*FVI* photograph by William I. Oliver.)

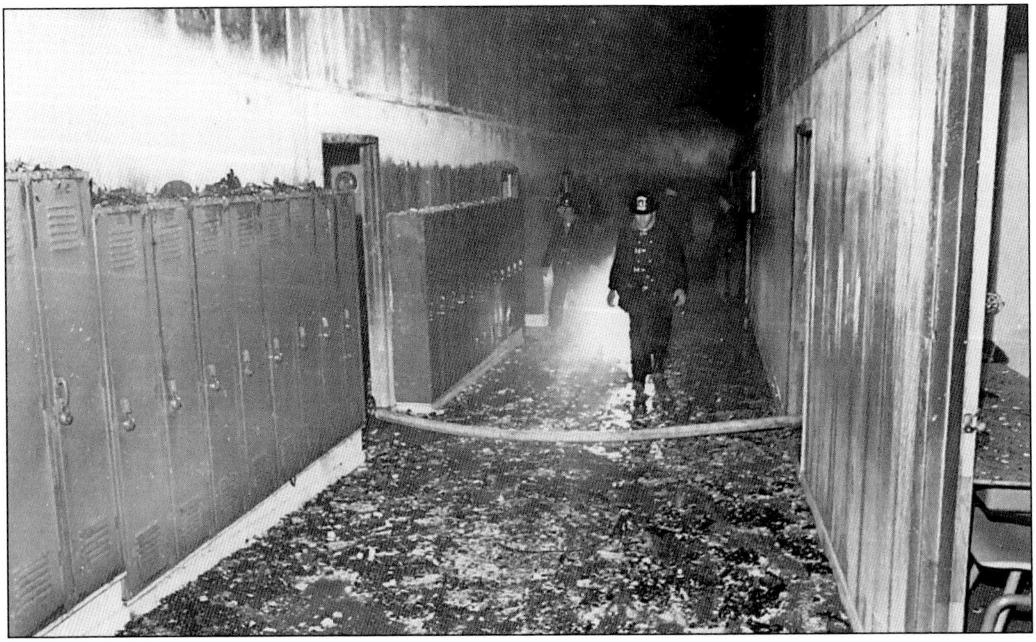

Broughton High School, February 20, 1969. This fire at 723 St. Mary's Street in Raleigh heavily damaged the old auditorium section of the 40-year-old school. Firefighters arrived within three minutes of the 2 p.m. alarm and quickly controlled the blaze. No injuries were reported to any of the 1,000 students or faculty members. (*N&O* photograph by Frank Urben.)

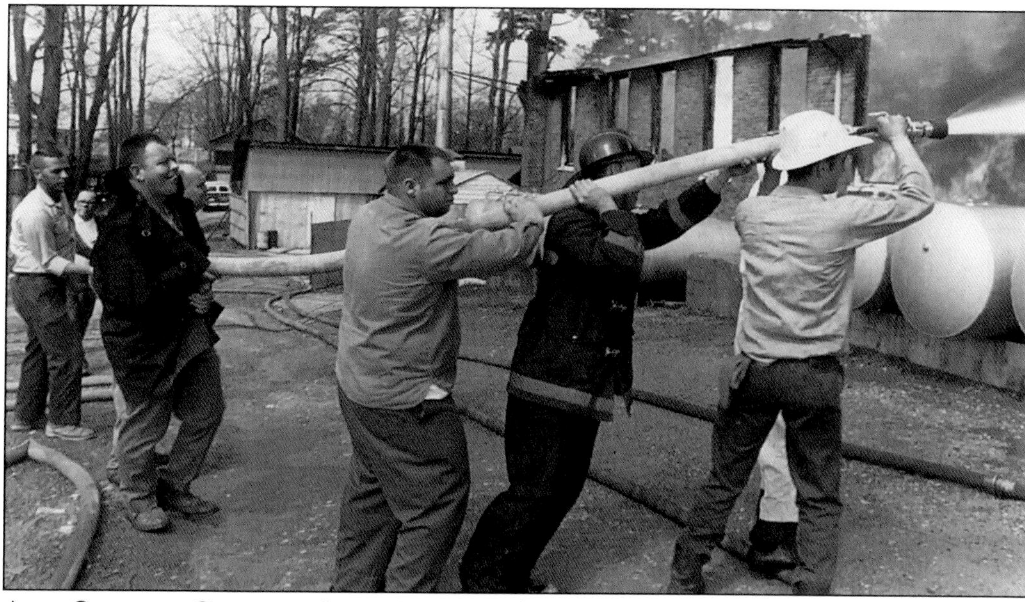

APEX CHEMICAL COMPANY, APRIL 6, 1970. The town's largest fire in decades started in the basement of this 43-year-old company on N. Salem Street. Units from Apex, Cary, and Yrac answered the 9 a.m. alarm. The fire, which shook the streets with occasional explosions, was brought under control in the early afternoon. (*N&O* photograph.)

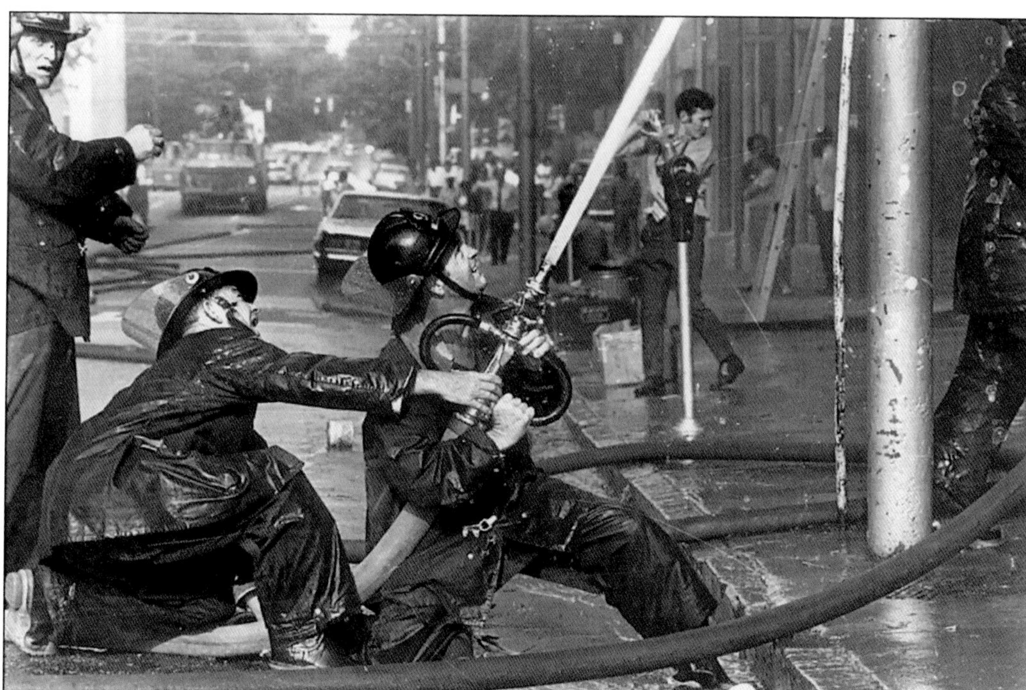

PEEBLES HOTEL, JUNE 24, 1970. The third floor of the downtown Raleigh landmark was destroyed by this five-alarm fire. Smoke and water extensively damaged the rest of the 1921 building, located at 122 E. Hargett Street. Orlin Wilder, Lawson Mitchell, and James Atkinson are pictured from left to right. (*N&O* photograph.)

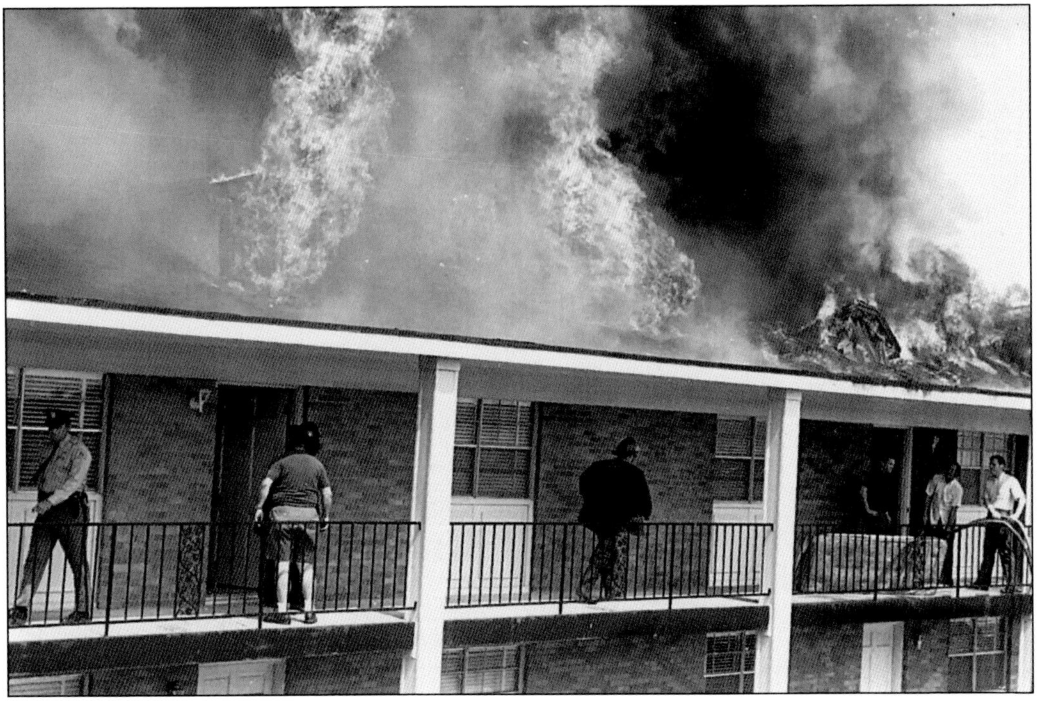

North Valley Apartments, September 6, 1970. Three second-floor apartments at 4343 Lassiter Mill Road were heavily damaged by this spectacular Sunday blaze. It was reported just after noon by a tenant taking a shower, who looked up and saw the ceiling on fire. The management offices were also damaged. (*N&O* photographs.)

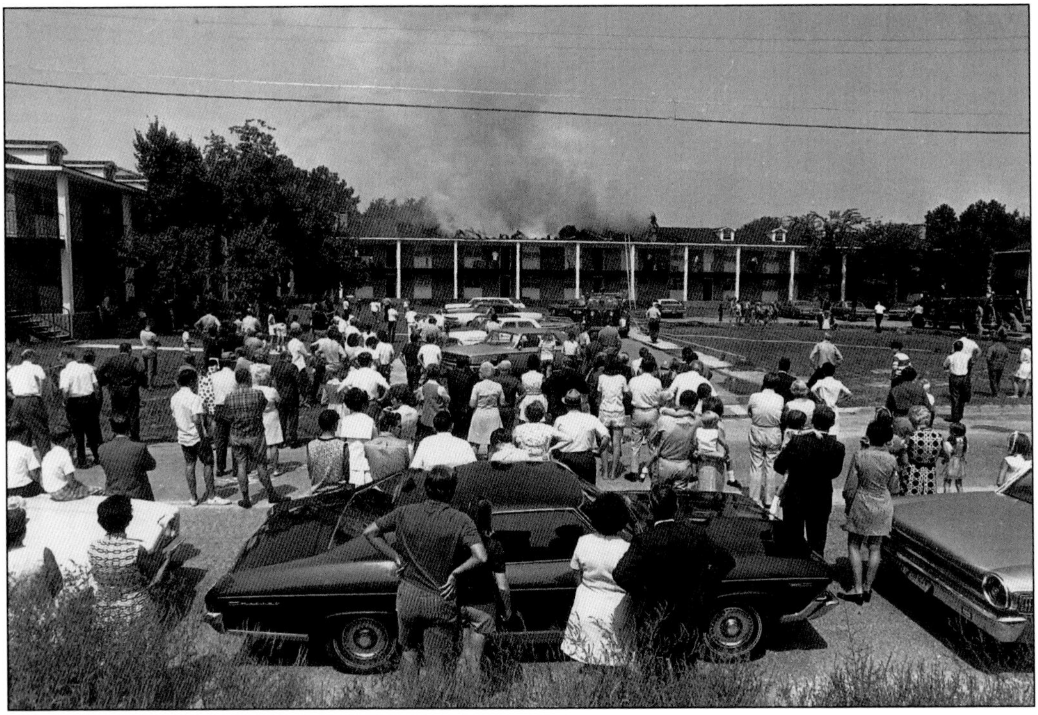

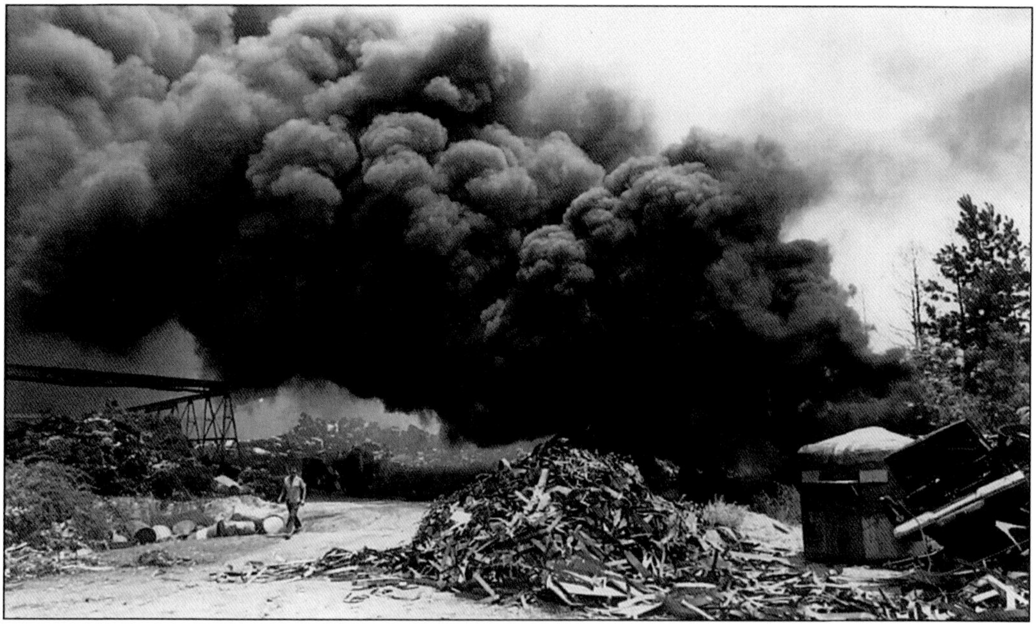

K&L Scrap Yard, July 20, 1970. A black cloud blew over Raleigh from this half-acre fire at K&L Scrap Service on Old Garner Road. Most of the smoke came from a pile of tires, though several stacks of junked cars also burned. The main building was not damaged. (*N&O* photograph by Bob Webb.)

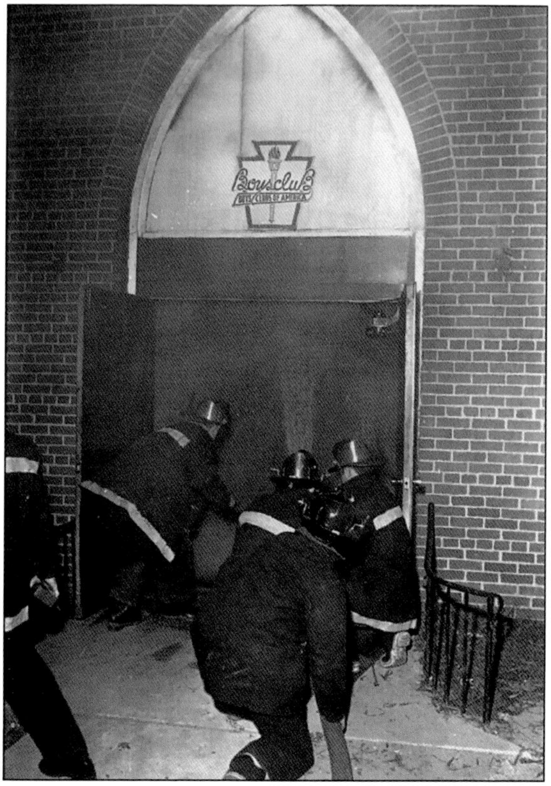

Boys Club, December 22, 1971. This Quonset building at 1109 New Bern Avenue was gutted by a 5:52 p.m. fire. The destroyed contents included three billiard tables, three ping pong tables, and four tumbling mats. The metal structure with a brick facade was owned by St. Joseph's Catholic Church and was leased to the Boys Club. Three other fires were reported that night and early morning in Raleigh and Wake County: a small fuel oil fire at Women's Prison in Raleigh, a house fire that the owner extinguished at 1621 Wilson Road near Cary, and a $10,000 fire at the historic Winstead Homeplace near Wendell. (*N&O* photograph.)

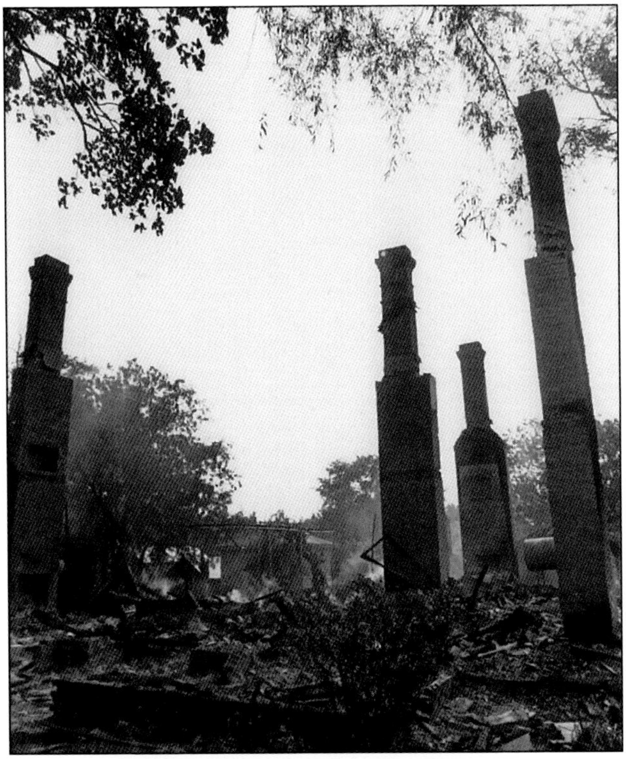

Walnut Hill, June 24, 1973. One of Wake County's oldest homes burned to the ground three weeks after owners announced its up coming use as a drug rehabilitation center. Vacant since June 1, the 16-room house in Shotwell was reported aflame at 12:54 a.m. Units from Knightdale, Wendell, and Garner responded and firefighters battled the blaze for several hours. The antebellum home was built around 1775 and originally had been given to the owner's family as a land grant from King George III. (*N&O* photograph by Steve Adams.)

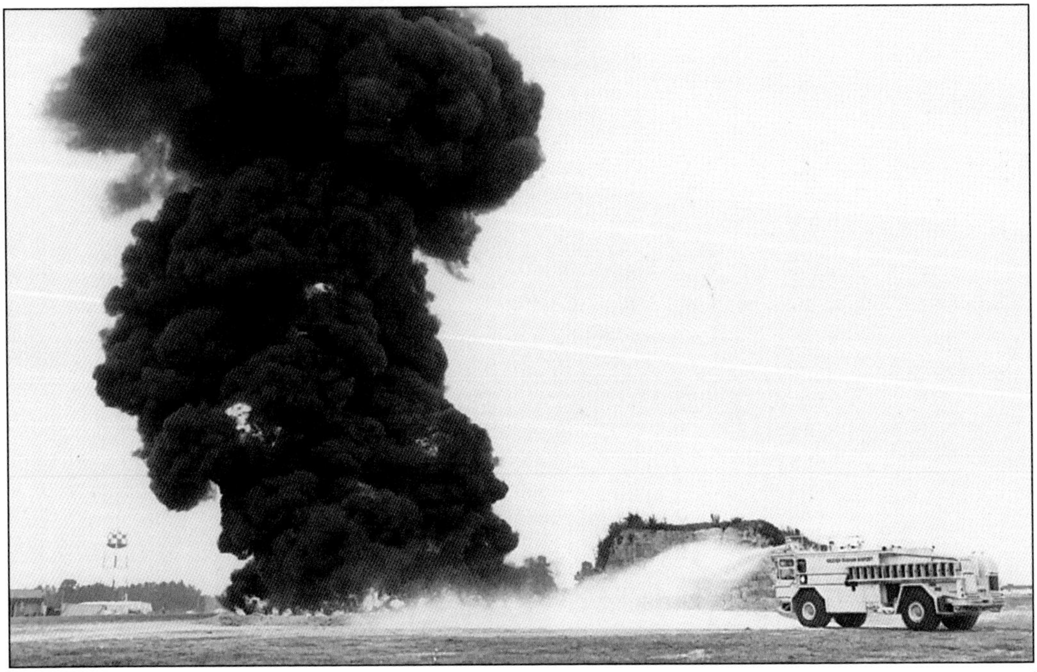

Raleigh-Durham Airport, August 9, 1973. Aviation fuel burns at a demonstration of the airport's newest crash truck. Powered by twin diesel engines, the 460-horsepower Walter CB3000 carried 3,000 gallons of water and 500 gallons of foam. The 33-ton behemoth cost $144,000. (*RT* photograph by Mike Clemmer.)

FATAL FIRE, OCTOBER 5, 1973. Blankets cover the bodies of a 31-year-old woman and her five-year-old-son, discovered apparently suffocated at 5717 Sharon Street in Garner. The fire was discovered by an off-duty Raleigh fireman who was painting a nearby house. After shouting to a neighbor to call the fire department, he found a ladder, ran to the house, and broke a window in the middle bedroom. Smoke and heat prevented his entry, and he could not see the bodies of the victims below the window. The mother's body was found lying over her son, as if shielding him from the smoke. The door to their room had not been closed. (*RT* photograph by Steve Murray.)

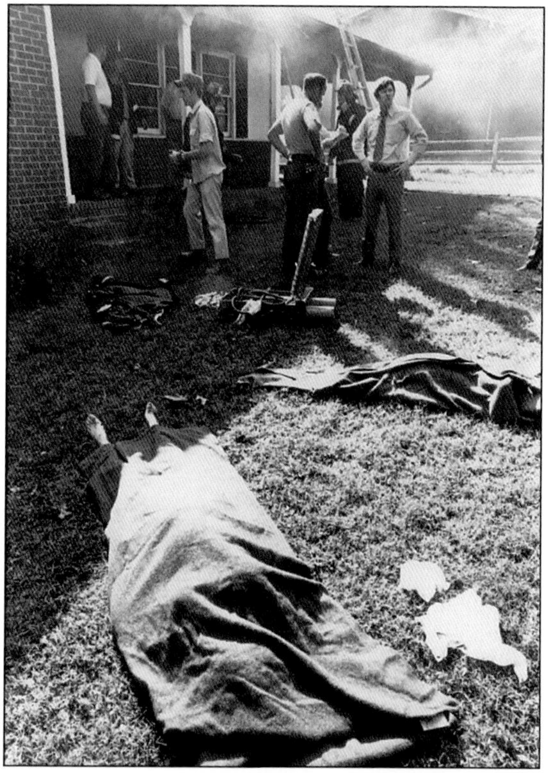

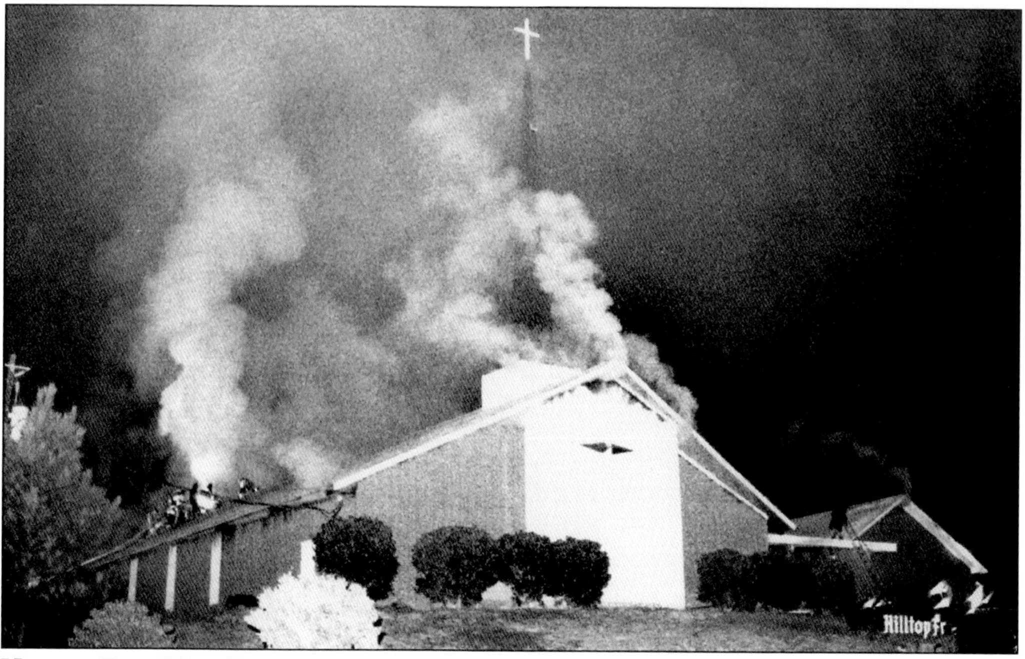

HILLTOP FREE WILL BAPTIST CHURCH, FEBRUARY 14, 1975. This Friday-night fire started in a secretary's office and destroyed most of the southern Wake County church on Highway 401. Units from Fairview, Fuquay-Varina, Garner, Holly Springs, and Swift Creek responded. The spectacle attracted about 2,000 spectators by midnight. (*N&O* photograph by Steve Murray.)

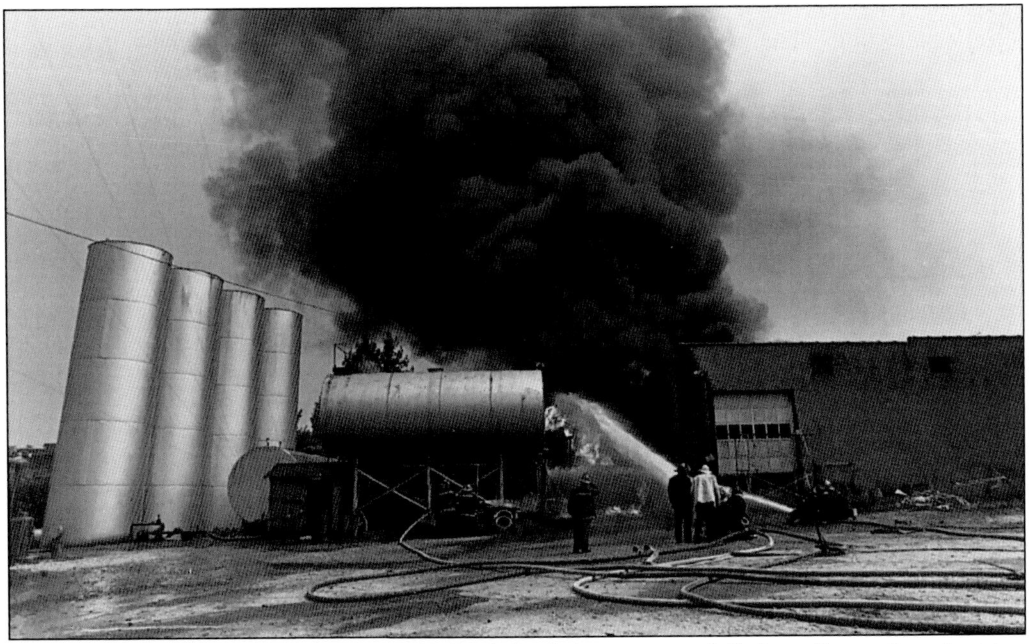

WHITE OIL COMPANY, JULY 10, 1975. Raleigh firefighters successfully kept flames from igniting 100,000 gallons of fuel oil at 1115 W. Lenoir Street. Fearing an explosion, officials evacuated several homes in the area. A crash truck from the airport was also summoned. Because of the vehicle's height, personnel were posted on the roof as the 1973 Walter CB3000 responded along Western Boulevard. They raised power lines, traffic signals, and other obstructions using pike poles. (*N&O* photograph by Steve Murray.)

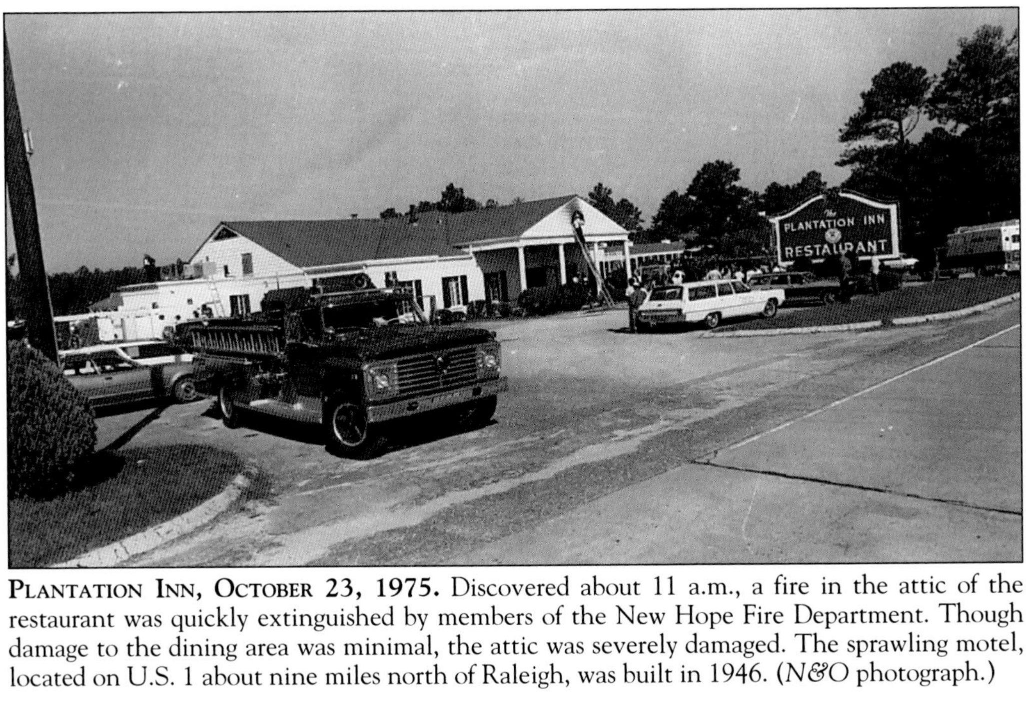

PLANTATION INN, OCTOBER 23, 1975. Discovered about 11 a.m., a fire in the attic of the restaurant was quickly extinguished by members of the New Hope Fire Department. Though damage to the dining area was minimal, the attic was severely damaged. The sprawling motel, located on U.S. 1 about nine miles north of Raleigh, was built in 1946. (*N&O* photograph.)

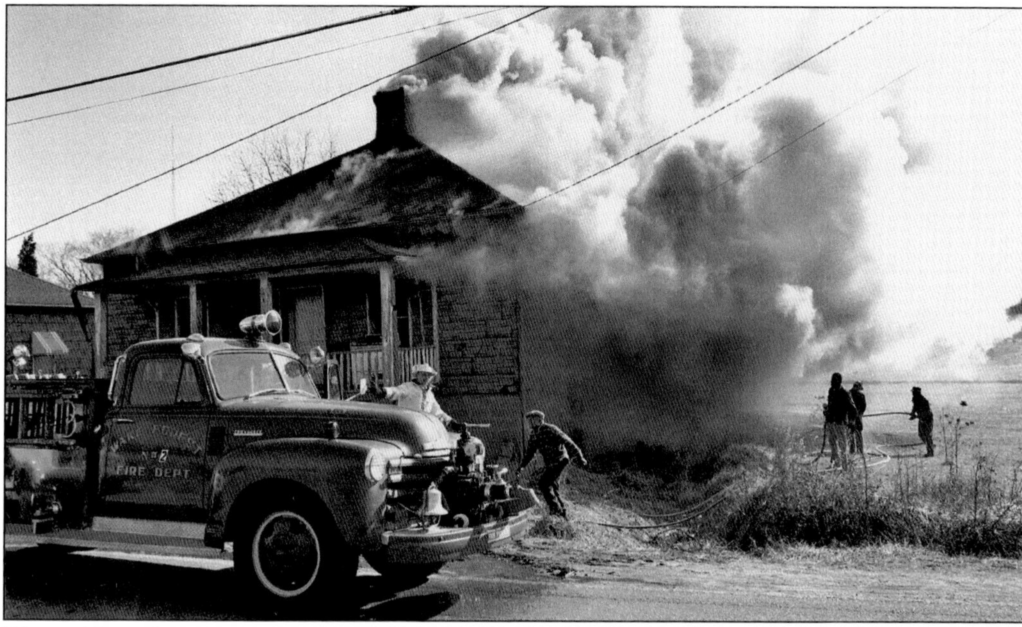

House Fire, January 17, 1977. Wake Forest firefighters battled sub-freezing temperatures at this house fire on N. White Street. The occupants, a family of 10, lost all of their possessions. The fire department was recalled twice to the scene, first for flames sighted in an area around the chimney and again after furniture inside the house re-ignited. (*WW* photograph by Bob Allen.)

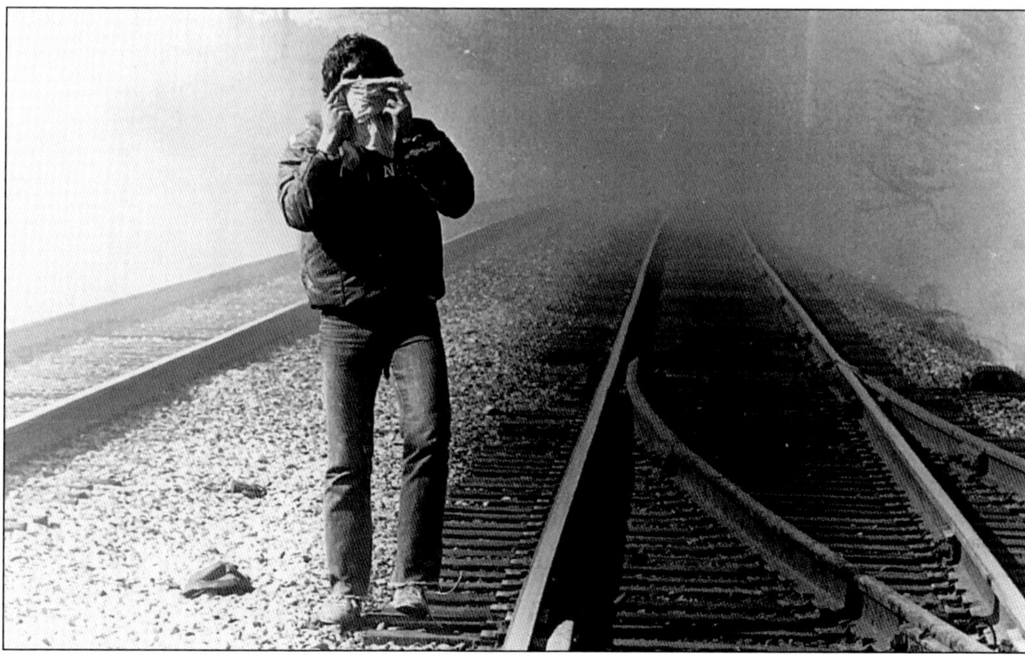

Brush Fire, February 5, 1977. An unidentified North Carolina State University student walks along the railroad tracks near Sullivan Hall where a half-mile stretch of brush burned, threatening residential storage sheds on Stanhope Avenue. Flames also scorched cars parked next to the baseball field. (*N&O* photograph by Scott Stewart.)

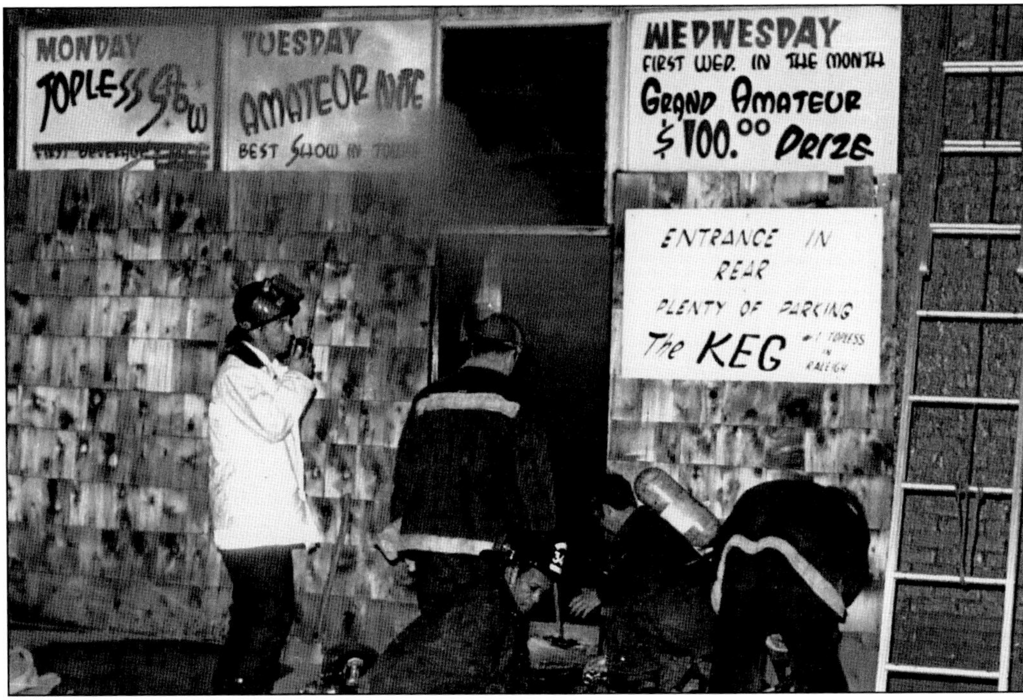

THE KEG, MARCH 12, 1977. Believed to have started in a trash can behind the building, an early Saturday-morning fire heavily damaged this tavern at 3104 Hillsborough Street. An adjoining game room also received smoke damage. (RFD.)

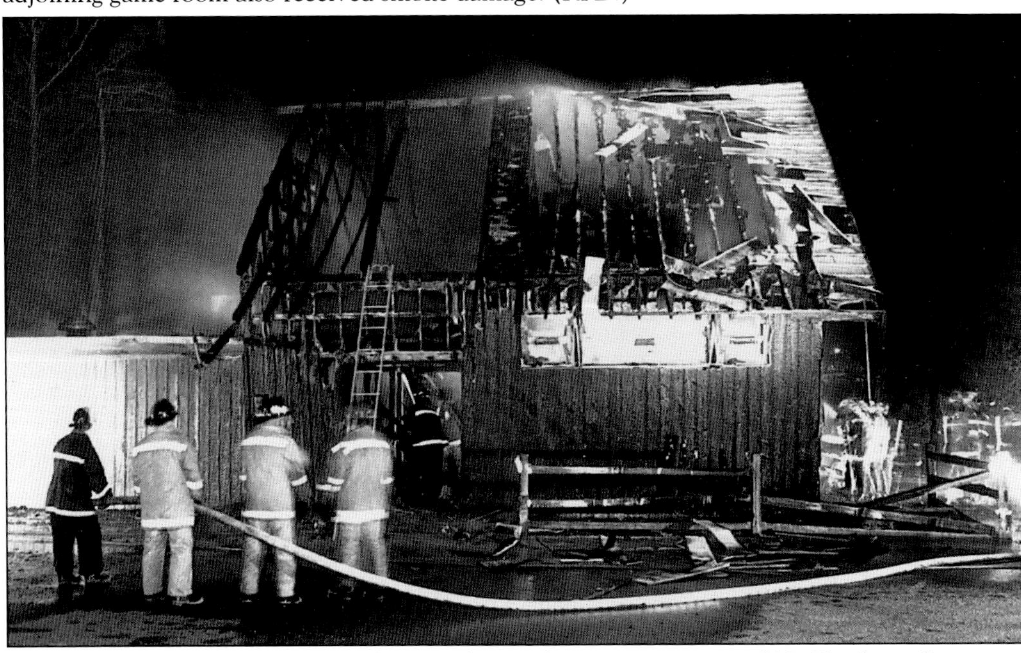

COOK OUT, MARCH 23, 1977. This wooden A-frame restaurant at 500 Chatham Street was fully involved when Cary firefighters arrived at 3:29 a.m., one minute after the alarm was received. Flames were fed by more than a dozen cans of paint stored on the second level of the structure. (*RT* photograph by Greg Stewart.)

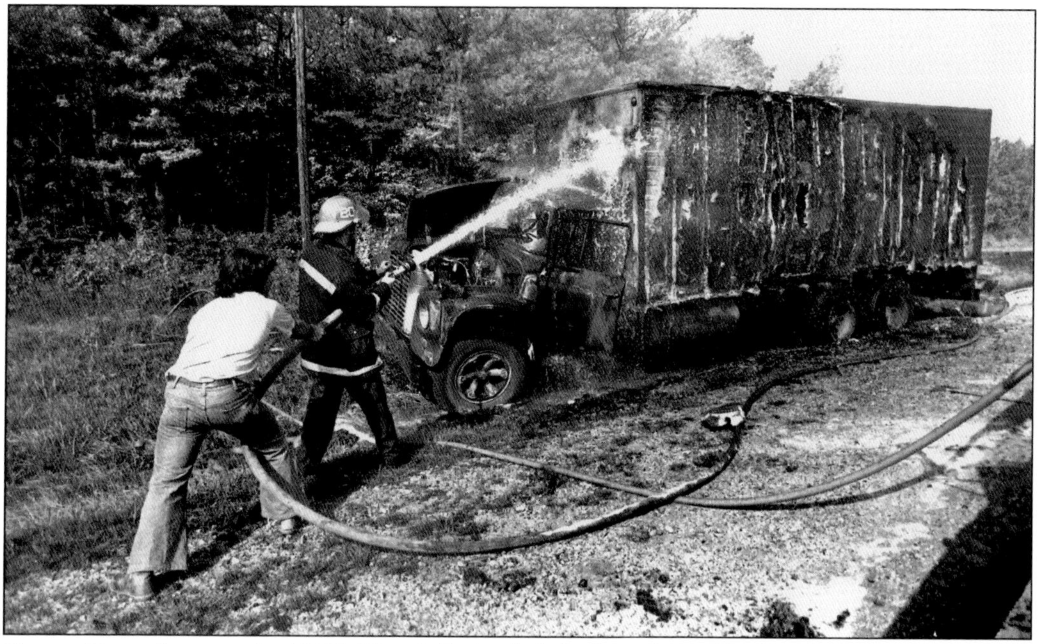

Truck Fire, August 8, 1977. This fully loaded food truck caught fire on Highway 70 West near the airport exit. A crash truck from the airport responded, along with two Durham Highway units. Smoke from the 80-foot flames was visible for five miles. (*RT* photograph by Steve Murray.)

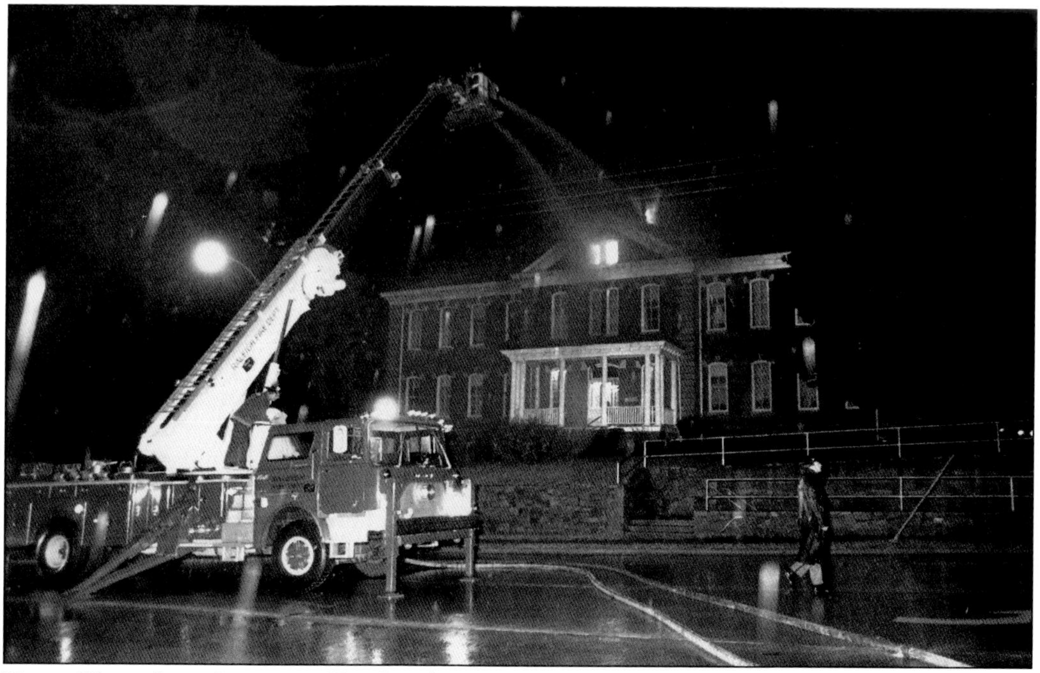

Tyler Hall, July 31, 1978. Truck 1 directs a master stream into the historic Shaw University building. Lightning started the 8:54 p.m. fire. Built in 1910, Tyler Hall originally housed Leonard Hospital, part of the college's medical school that closed in 1918. (*N&O* photograph.)

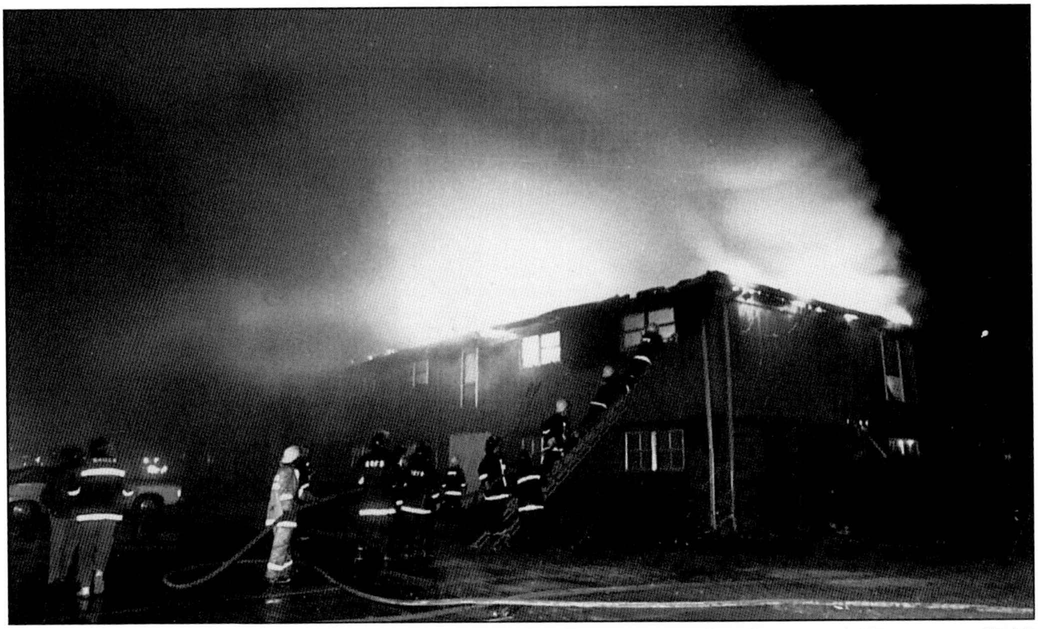

Fairway Apartments, December 30, 1978. This fast-moving fire destroyed 12 units at 5119 New Hope Road, a half-mile north of the Raleigh city limits. About 100 firefighters from several departments and 10 trucks from New Hope, Raleigh, and Rolesville battled the blaze. Damage was estimated at about $250,000. (*N&O* photograph by Jackson Hill.)

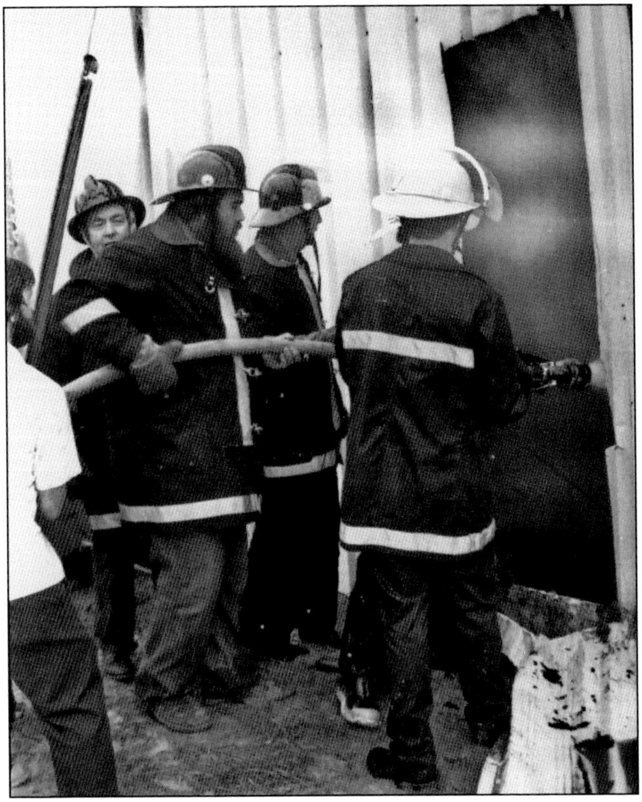

Bill's Outlet, May 20, 1980. Volunteers from six Wake and Johnston County fire departments assisted Wendell firefighters at this warehouse owned by Bill's Outlet on E. Wilson Road. The Tuesday-night blaze was believed intentional, and a former employee was charged with arson. The metal-frame warehouse was filled with kitchen cabinets. A passing motorist discovered the 6 p.m. fire. It was brought under control within two hours, though personnel remained on the scene into the night to guard against rekindling. Several firefighters were also treated for smoke inhalation and exhaustion. (*GLF* photograph by David Roberson.)

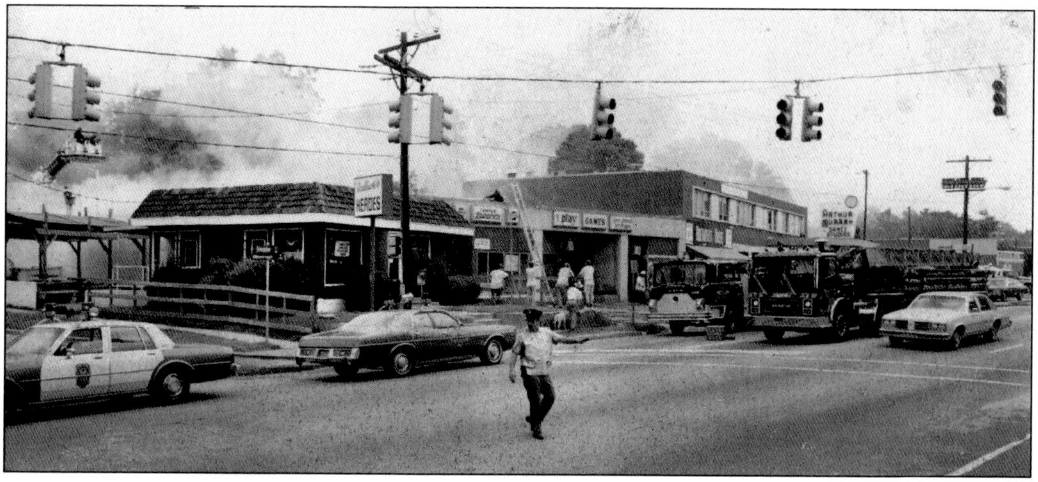

LAUNDROMAT, AUGUST 8, 1981. Raleigh firefighters responded twice that day to the Campus Launderette at 2114 Hillsborough Street. The first call came at 11:30 a.m. after lint caught fire above one of the dryers. Twenty minutes later, the fire department returned to the scene. The roof was burning and flames were spreading to the I Play Games arcade and a North Carolina State University Agricultural Extension Service workshop. Firefighters extinguished the second blaze in about a half-hour. (RFD.)

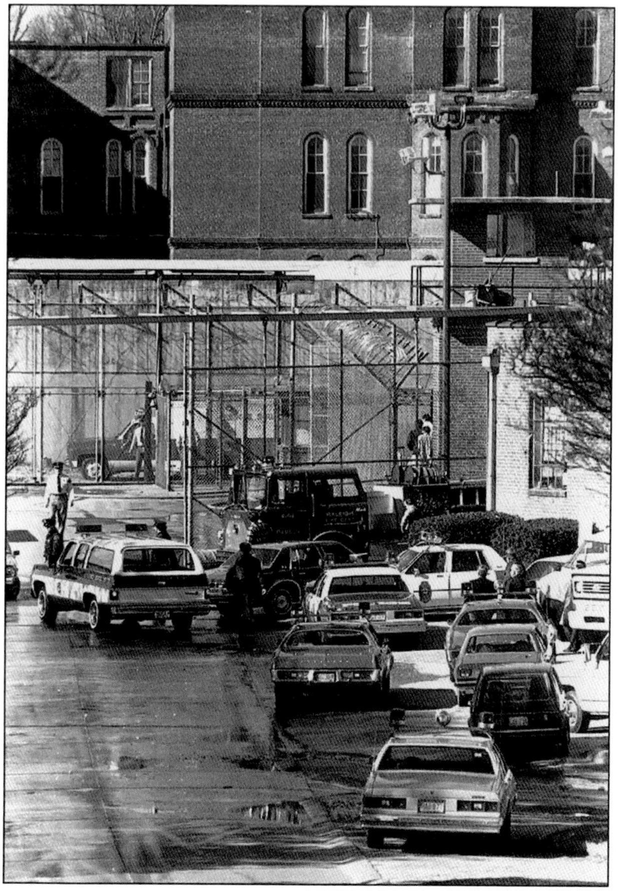

CENTRAL PRISON, JANUARY 5, 1982. One hundred inmates were evacuated from the Correction Enterprises building after a fire started in the corner of the sign plant. Reported at 11:40 a.m., arriving firefighters waited outside the gate until the inmates, guards, and Correctional Enterprises employees were moved into the prison yard. The fire, which destroyed between $80,000 and $100,000 worth of printing materials, was extinguished about 12:15 p.m. No injuries or security problems were reported. (*N&O* photograph.)

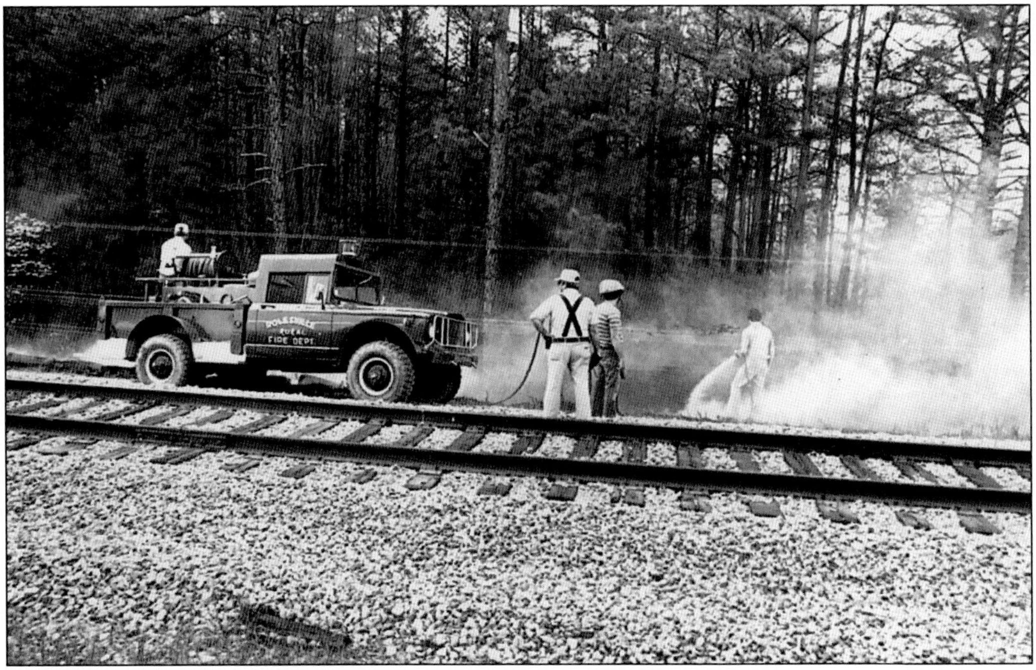

WOODS FIRE, APRIL 14, 1982. The Rolesville Fire Department works at a woods fire near Wendell. About 25 acres burned for five hours before being extinguished by firefighters from five departments. (*N&O* photograph.)

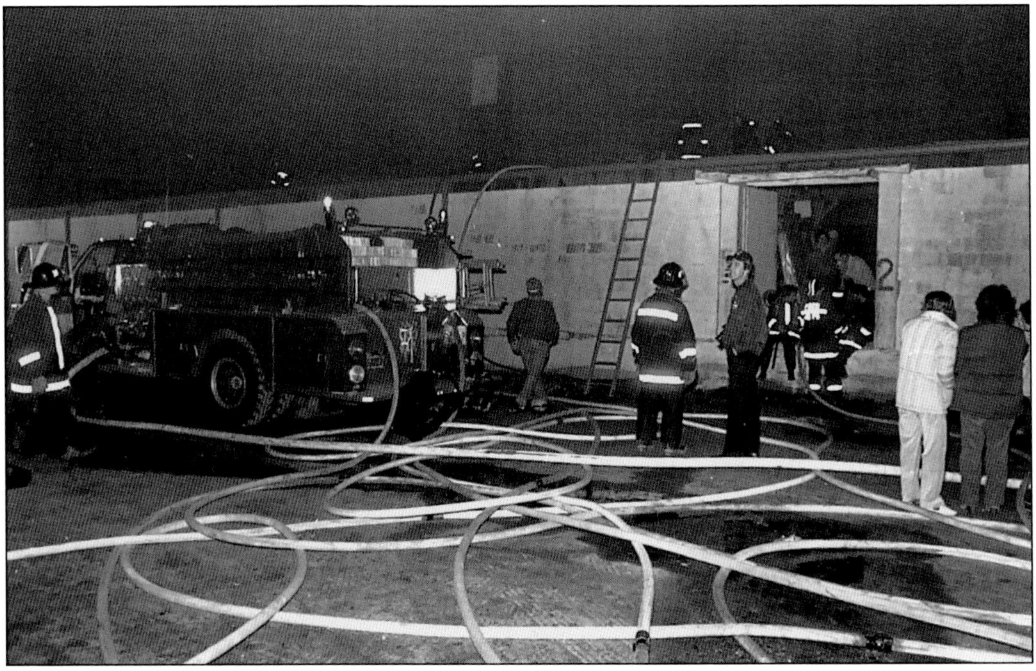

WHITLEY FURNITURE GALLERIES, JANUARY 8, 1983. This 10:45 p.m. fire started in the snack room of the furniture store at 100 W. Vance Street in downtown Zebulon. Firefighters from Wendell, Knightdale, and Hopkins helped the town's fire departments bring the blaze under control by 1:30 a.m. (*N&O* photograph by Robert Willett.)

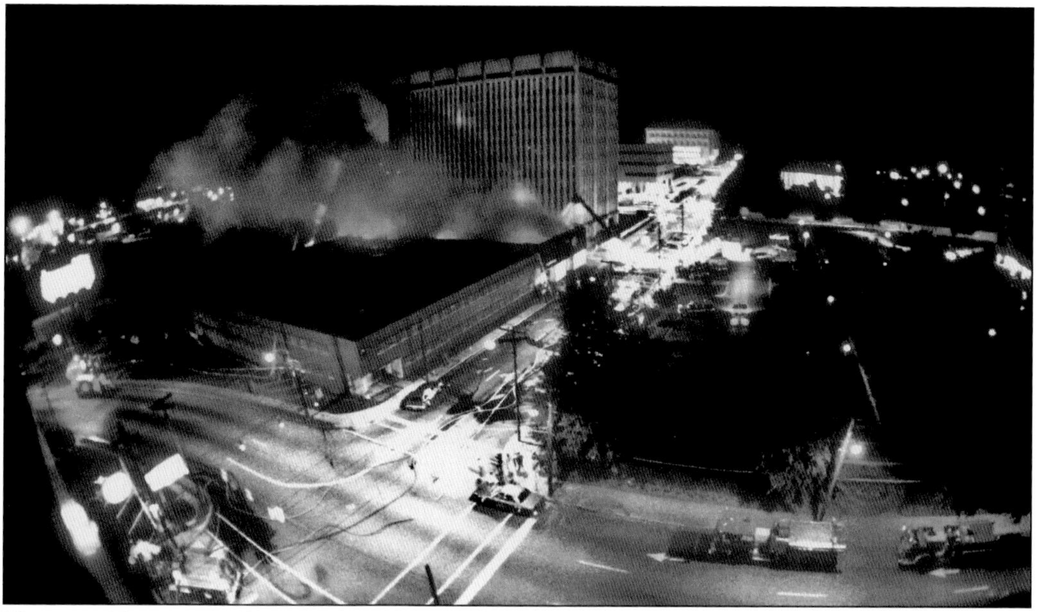

Howard Building, October 3, 1984. Blamed on a burning match in a cardboard box, a stubborn six-hour fire gutted this turn-of-the-century building at 112 E. Lane Street. Nearly 100 firefighters, working in shifts of 40, battled the 4:15 p.m. blaze. Volunteers from the Six Forks Fire Department assisted with refilling air tanks as more than 100 tanks were depleted by the end of the evening. This photograph looks east down Lane Street from McDowell Street. (Gary Knight/CCBI photograph.)

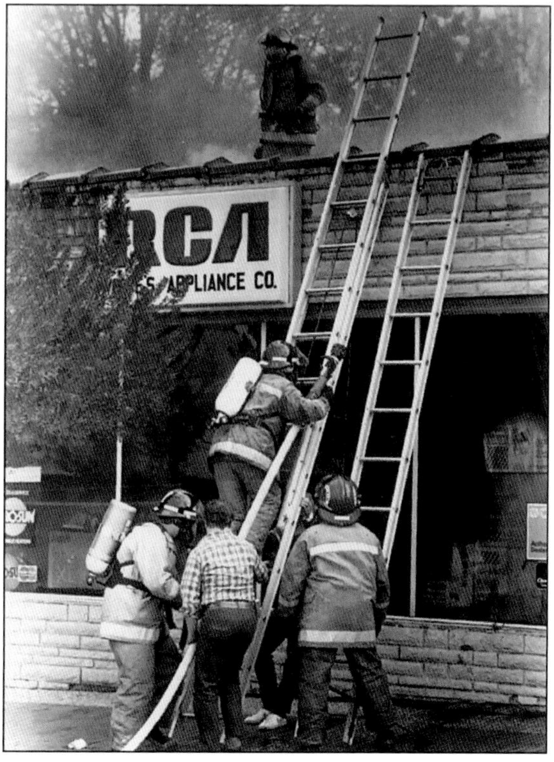

Wolfe's Appliance and Service, November 25, 1984. Smoldering insulation produced thick smoke for over an hour at this Sunday-afternoon fire in downtown Cary. A passing police officer discovered the same fire at 140 E. Chatham Street about 3:05 p.m. Though brought under control quickly by some 25 firefighters, insulation in the building continued to burn. Police rerouted traffic around the area for about two hours. Firefighters remained on the scene until 9:30 p.m. The 40-year-old building also housed Rich's Style Shop and Jordan's Jewelers, with fire walls separating the businesses. Fire investigators determined that the blazed started in a waste basket behind a counter in the rear of the store. (*N&O* photograph by Whit Elfner.)

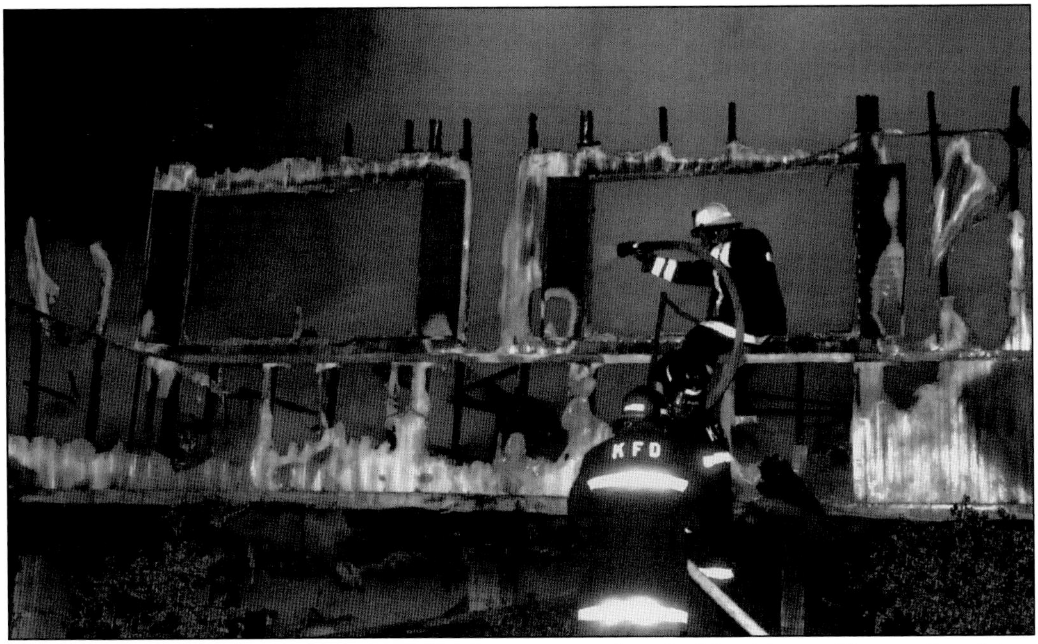

Mobile Home, November 9, 1985. Little remains of a doublewide on Horton Road just south of Jackass Road in Knightdale. Reported just after 8 p.m., the trailer was destroyed along with an outdoor equipment shed. The family's dog was also killed. (*GLF* photograph by Betty Patterson.)

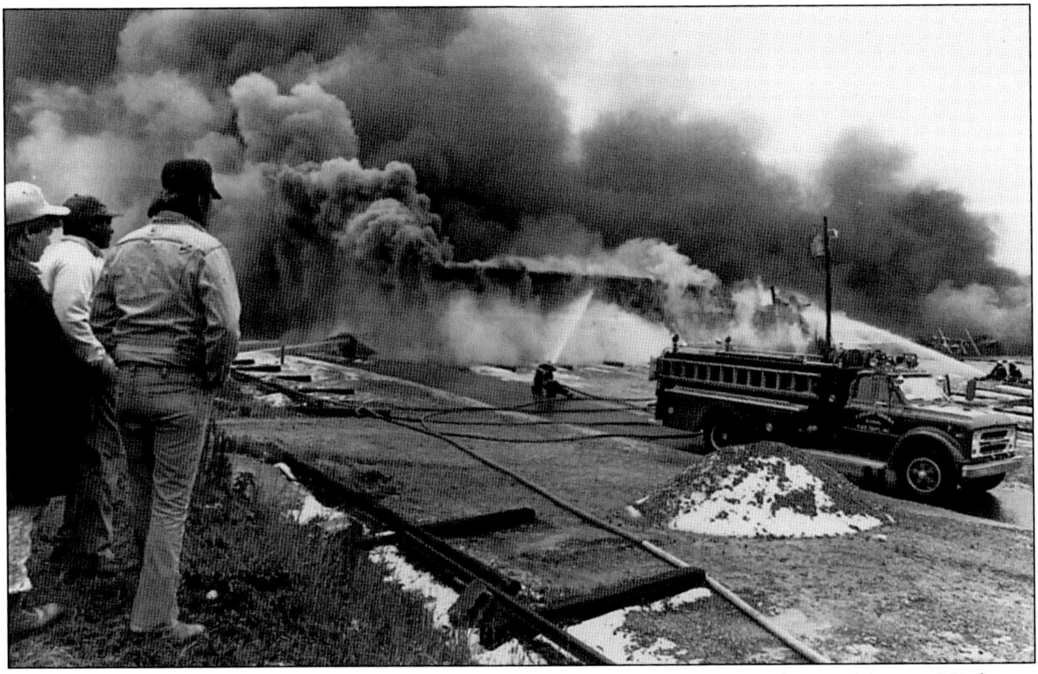

Fuquay-Varina Flea Market, December 11, 1989. Located on Highway 55 near Highway 401, the Carolina Gold Leaf flea market was destroyed by this Monday fire that started just before noon. Flames gutted the pinewood-and-metal building that was filled with about $1 million in merchandise. (*N&O* photograph.)

Two

EQUIPMENT

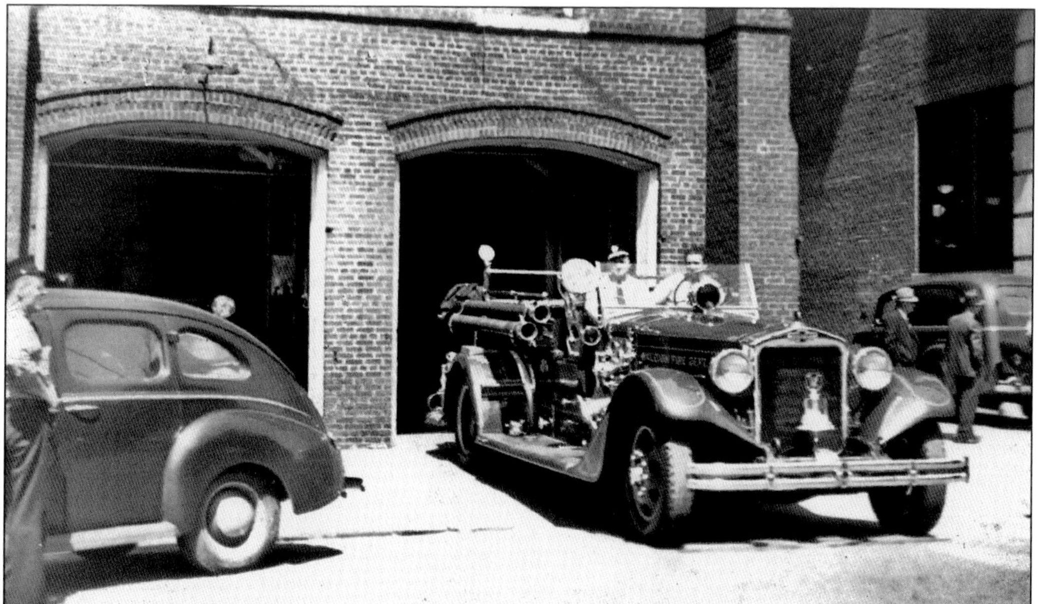

ENGINE 1, C. 1940. Pictured in front of Station 1 at 112 W. Morgan Street is the Raleigh Fire Department's lone American LaFrance 400 Series pumper. Delivered in 1936, the 1000-GPM fire engine remained in service until the late 1960s. (Courtesy of Mike Kimmeries.)

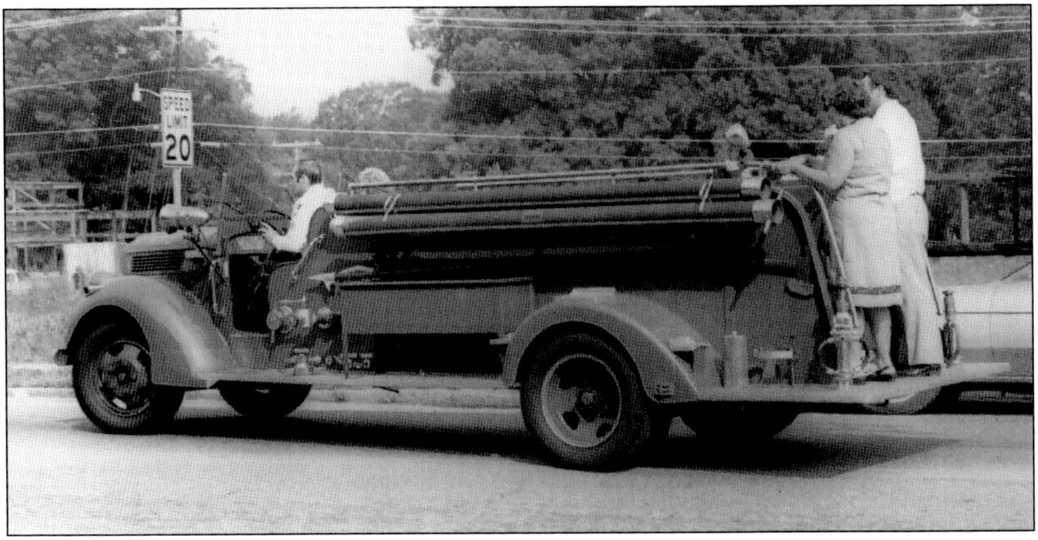

Town Fire Truck. This 1940 Ford/Hunter pumper was purchased new by Apex, one year after the volunteer fire department re-formed. The town's previous pumper was an American LaFrance Type 10 "triple-combination 500-gallon pumping, chemical engine, hose motor car" purchased in 1927. By the late 1930s there was no official fire department in Apex. Calls were answered by anyone who happened to hear the alarm and the old LaFrance was pulled to fires by the Clark Chevrolet Company wrecker. In 1938, 19 men met to address concerns about sufficient fire protection for the town. One year later, a new fire department was chartered and members converted a $35 dump truck into a temporary fire engine. (AFD.)

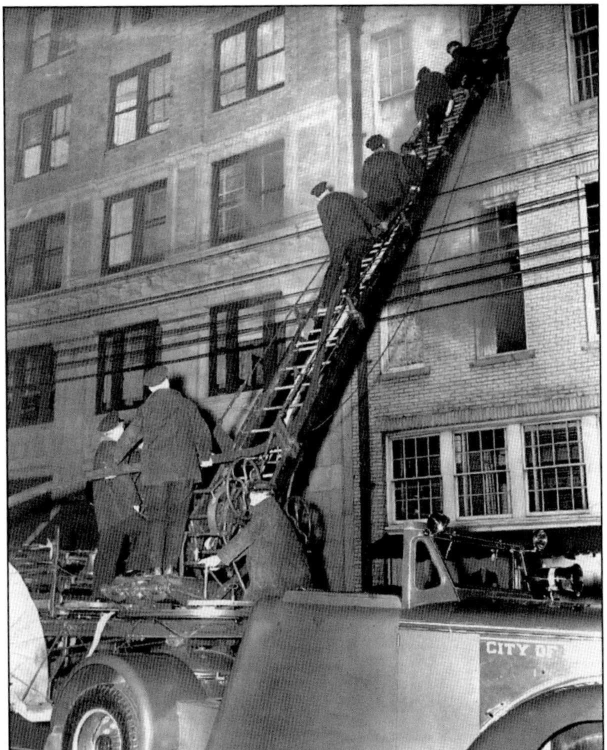

Ladder Truck. Raleigh's 75-foot American LaFrance aerial ladder was first delivered in 1916. The hand-cranked, spring-assisted, Type 17 wooden ladder was reconditioned in 1939 and outfitted with a new 500 Series tractor. An excerpt from the operator's manual on raising the ladder reads "The operator should now pull out the ladder lock. While standing in front of the operating mechanism, the gear reduction shift lever is placed in the neutral position. The friction brake handle is then pulled forward; this controls the speed of the ladder raise. Releasing the ladder locks to a horizontal position is next. Then, the foot trip lever is ready to be released which applies spring pressure to raise the ladder." (Courtesy of C.T. May.)

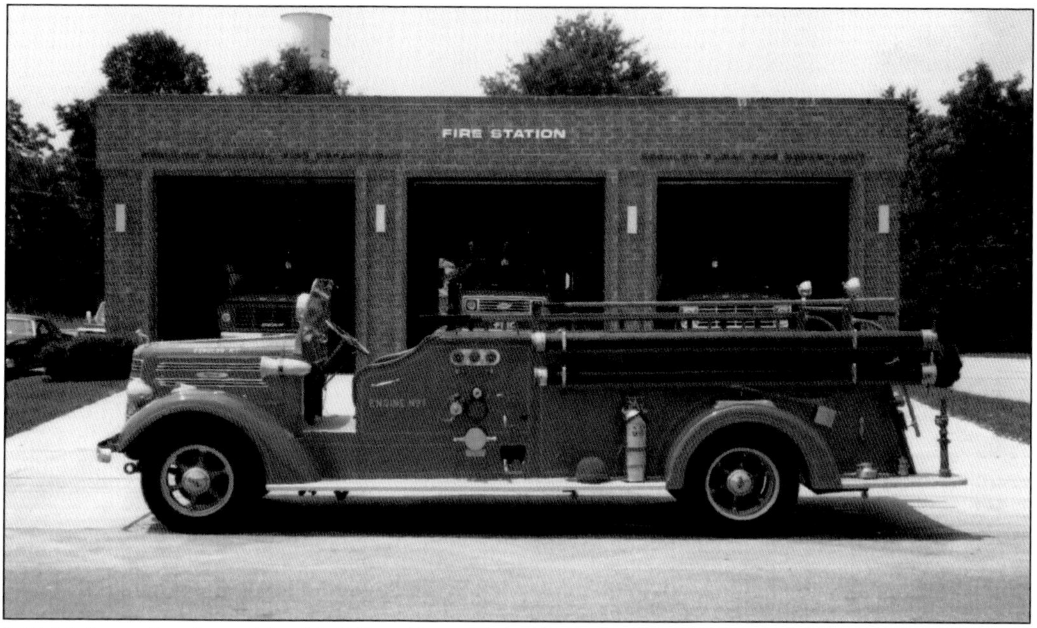

1945 Mack. Purchased for $7,150 in December 1946, this Mack Type 45 pumper was delivered new to the town of Zebulon. It was equipped with a 500-GPM pump and a 150-gallon tank. The open-cab engine replaced the town's first fire truck, a Ford Model A truck adapted by firefighters in 1928. The old Mack was sold for $750 to a town resident in 1977. He restored it for parade use. (ZFD.)

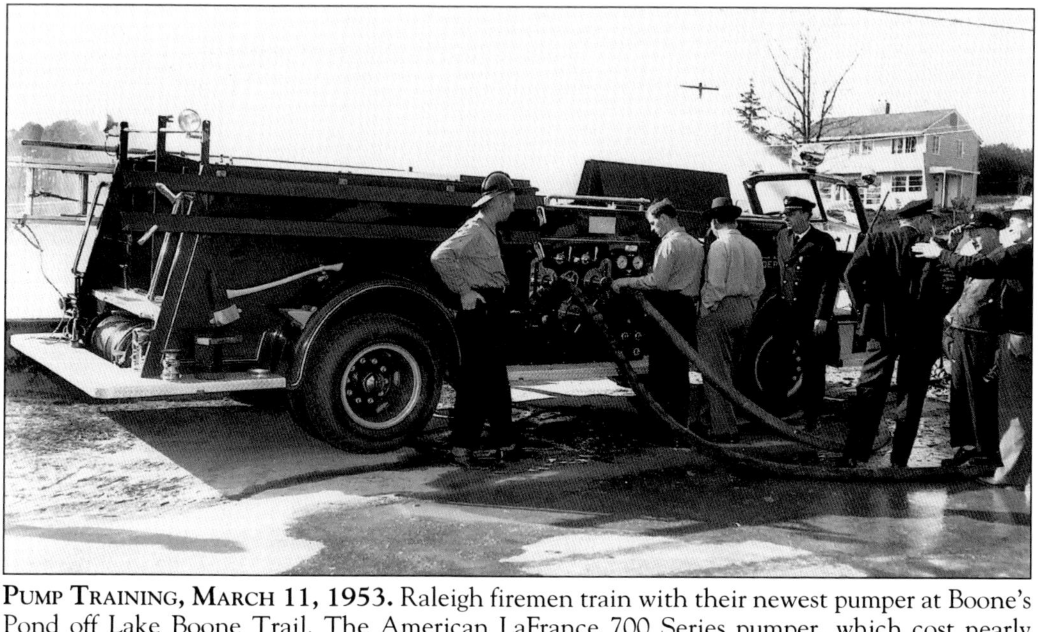

Pump Training, March 11, 1953. Raleigh firemen train with their newest pumper at Boone's Pond off Lake Boone Trail. The American LaFrance 700 Series pumper, which cost nearly $17,000, was delivered on March 8. Acceptance tests were conducted over the next several days. (*N&O* photograph.)

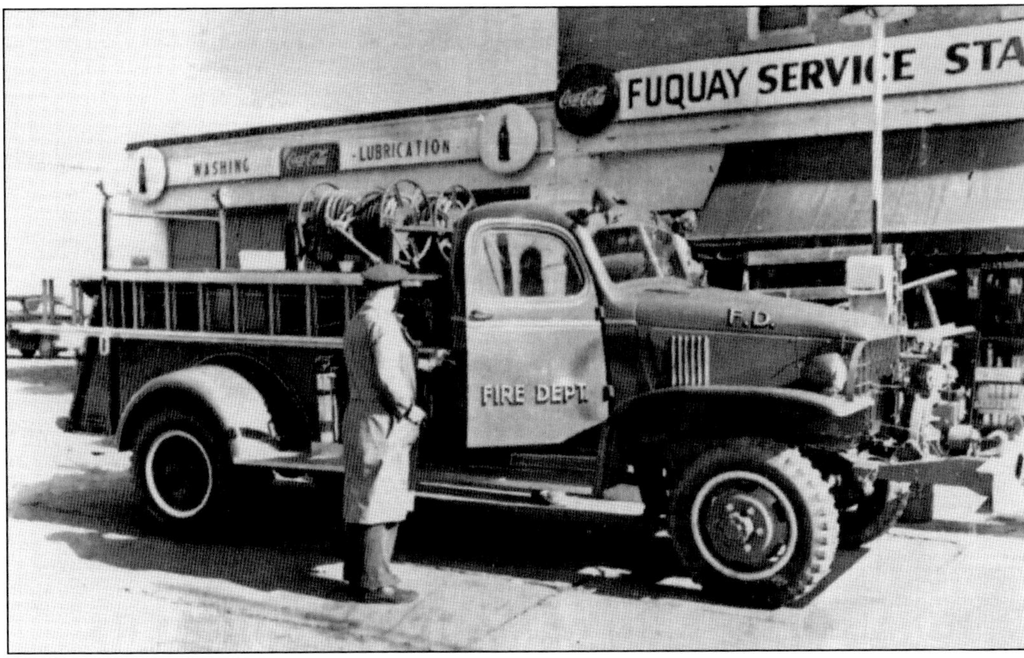

RURAL FIRE TRUCKS. The first fire engine of the Fuquay-Varina Rural Fire Department, pictured above, was a GMC 4x4 military surplus pumper loaned by the Howe Fire Equipment Company for several months until their new truck was ready. The rural fire department's second piece of apparatus was a tanker, pictured below, built by firefighters from an ex-military truck. Both were stored in a wooden shed behind the Fuquay Service Station at 104 S. Main Street. (FVFD.)

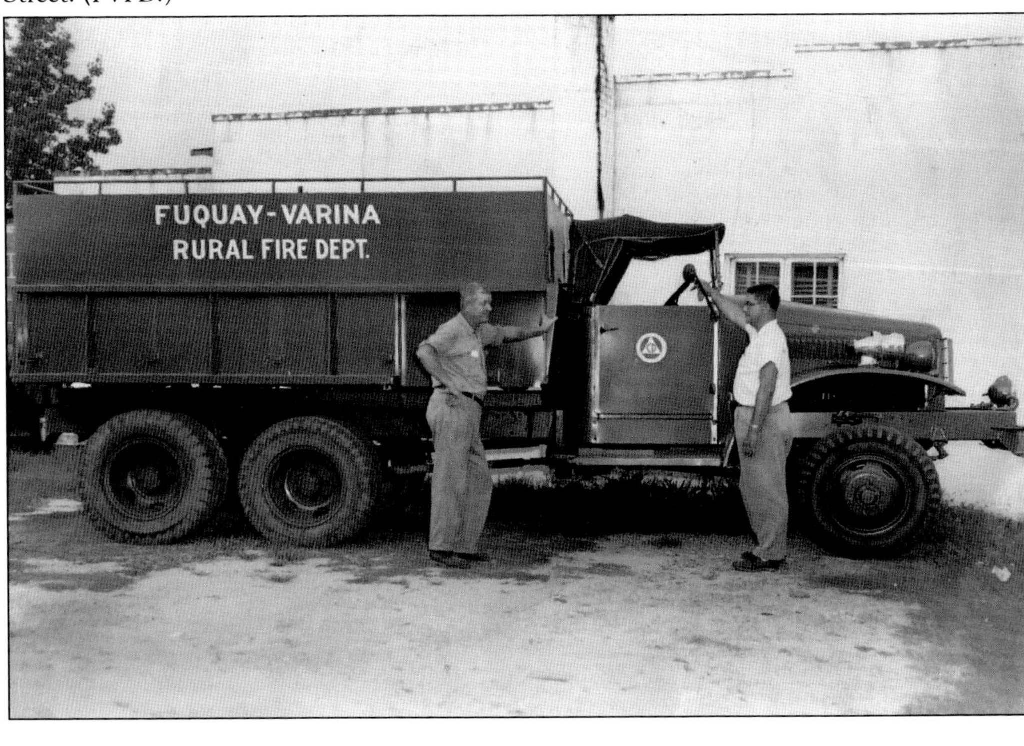

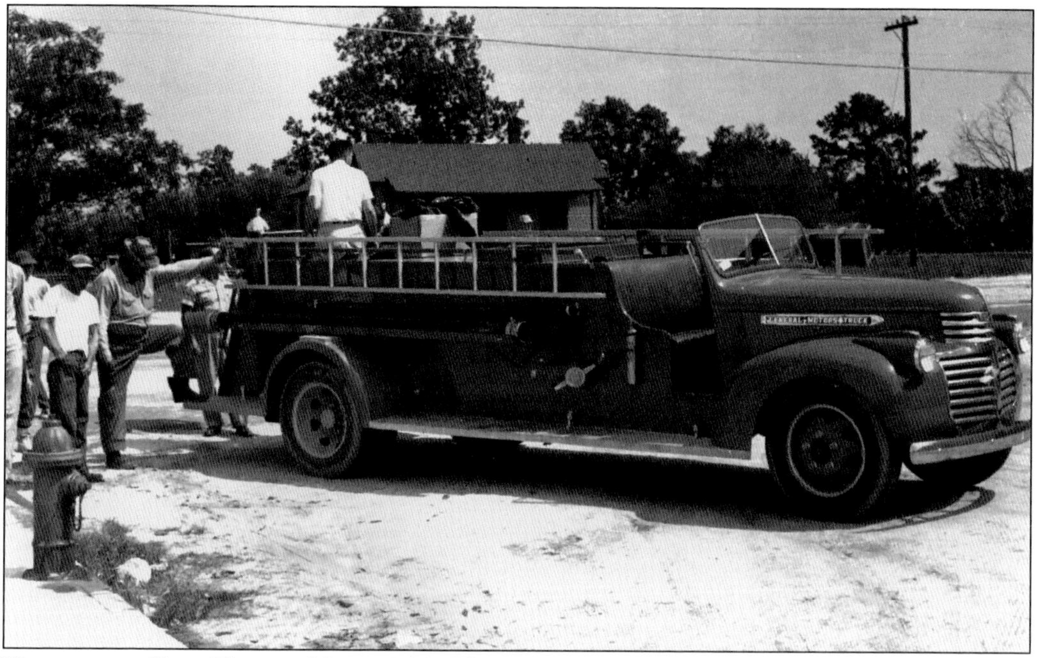

KNIGHTDALE FIRE TRUCK, MAY 1955. This 1942 GMC pumper was purchased from Navy surplus for $1,250. Equipped with a 500-GPM pump and 750-gallon tank, it was the second piece of apparatus for the two-year old fire department. Volunteers were using a one-and-a-half-ton town truck to carry hose, ladders, and other equipment. (John Stalvey photograph.)

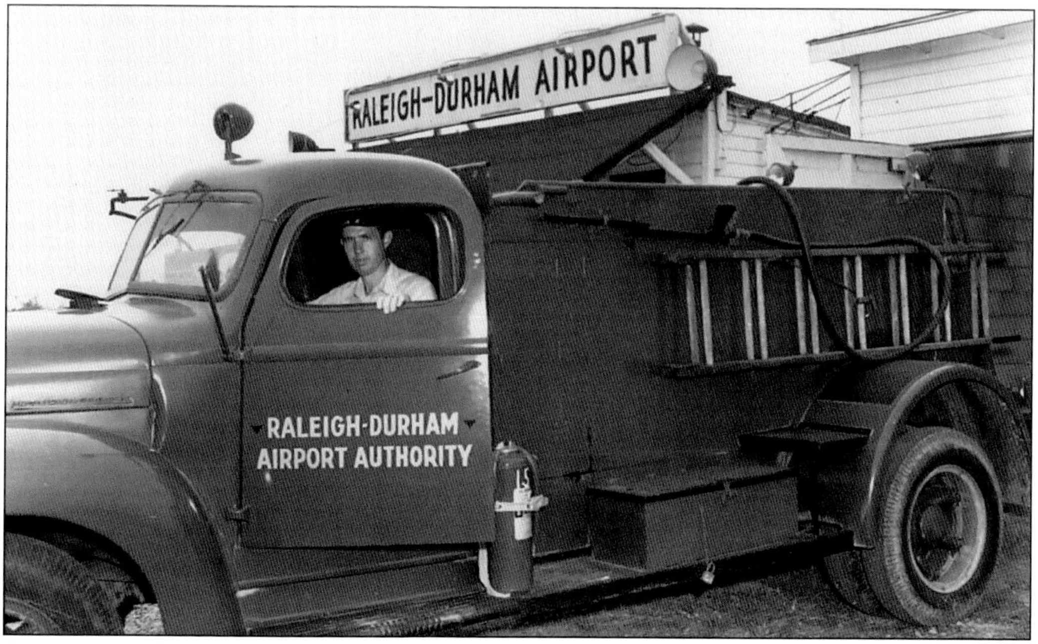

AIRPORT FIRE TRUCK, OCTOBER 1955. Bill Price of Eastern Air Lines ground operations, one of the airport's volunteer firefighters, poses in the cab of this 1946 International/Bean crash truck. It was one of two fire trucks stationed at the former Army air base. (*N&O* photograph.)

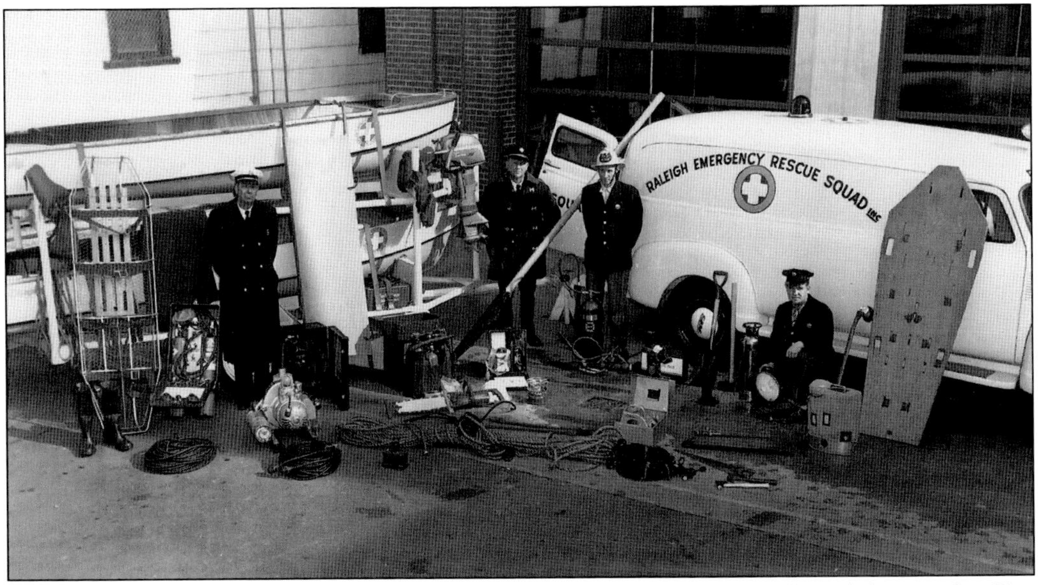

Raleigh Emergency Rescue Squad, December 1955. Pictured are, from left to right, Fire Chief Jack Keeter and squad members Royce Lassiter, Roma Wilder, and Walter Mabrey. The equipment on display includes a stokes stretcher, folding stretcher, resuscitator, gas mask, portable generator, cutting torch, chain saw, pry bar, block and tackle, telephones with cable, floodlight, fire extinguisher, canteen, and mummy board. (*N&O* photograph by R. W. Stephens.)

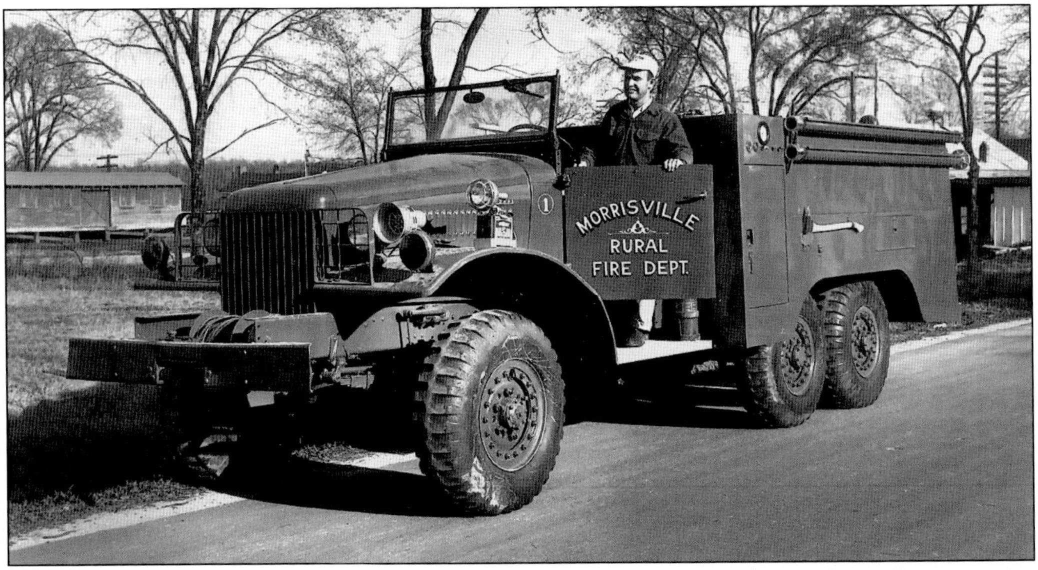

Morrisville Fire Truck, March 1956. Purchased last spring for $750, this military surplus tanker was converted into a fire engine by members of Morrisville's new volunteer fire department. They moved the 750-gallon tank farther back on the chassis and reworked the pumping system. Hoses, ladders, and other equipment were added at a cost of $1,500. The only outside work was $500 worth of skirting and $23 worth of painting. Per the minimum county requirement, the truck was supplied with 500 feet of one-and-a-half-inch hose and 400 feet of three-quarter-inch hose. It was first stored in a shed behind Jones and Sears Store on Cedar Street. Charter member Carl Light is pictured. (*RT* photograph.)

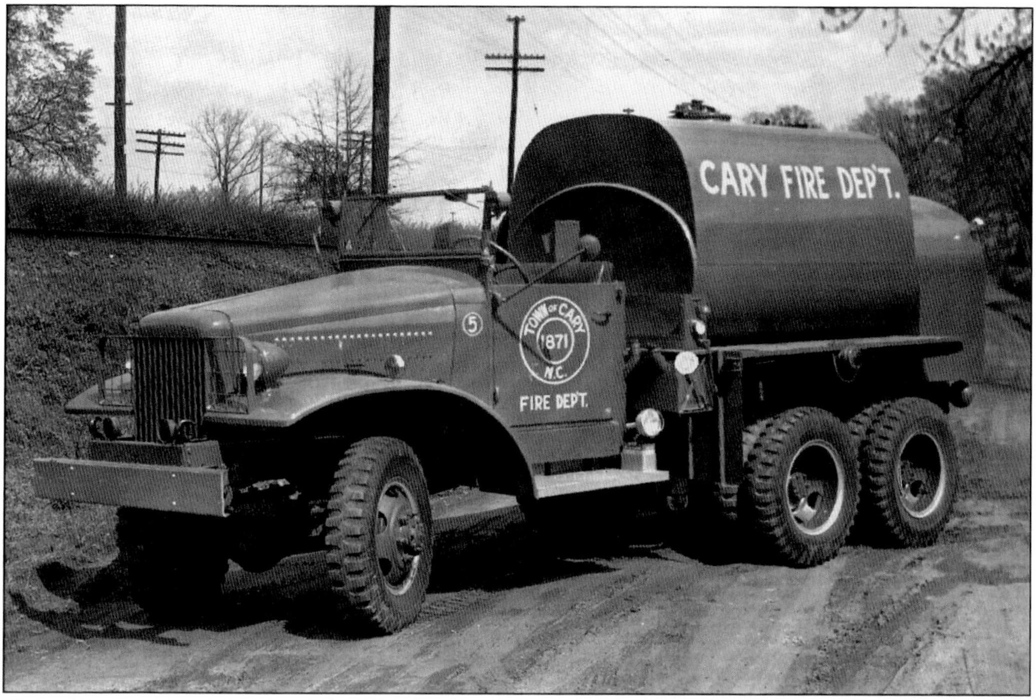

CARY TANKER, 1957. Firefighters rebuilt this ex-military truck into a tanker in 1956. The 1,500-gallon unit, pictured on Railroad Street opposite Station 2, was presented to the town at that year's Fireman's Day. (CFD.)

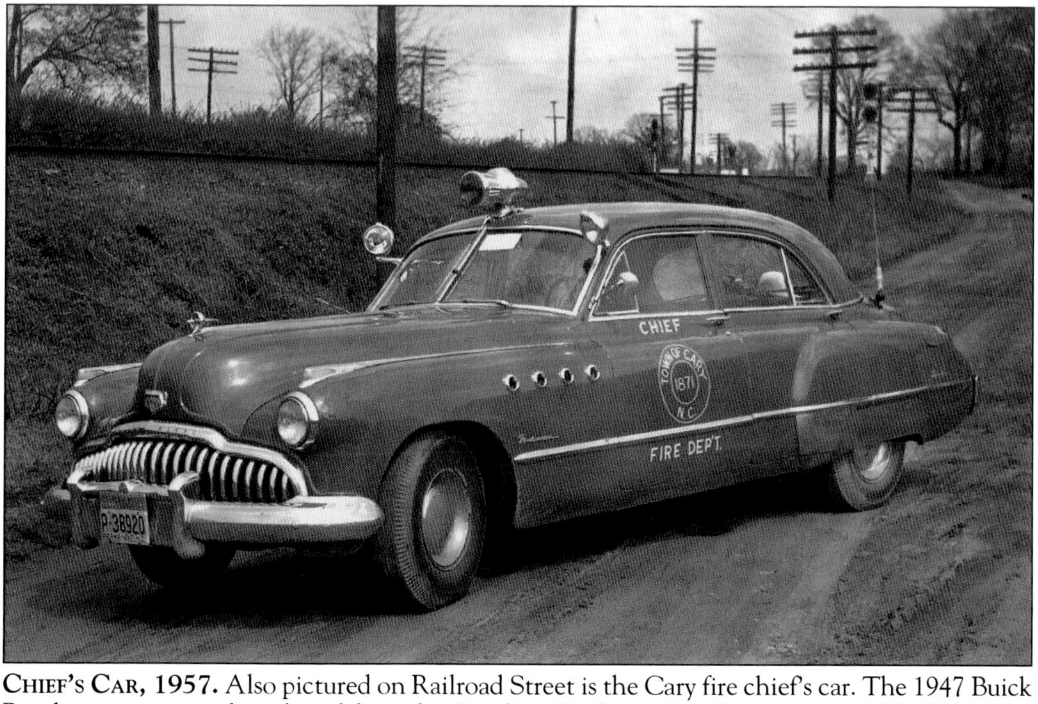

CHIEF'S CAR, 1957. Also pictured on Railroad Street is the Cary fire chief's car. The 1947 Buick Roadmaster was purchased used from the Carolina Trailways Bus Company in 1955. (CFD.)

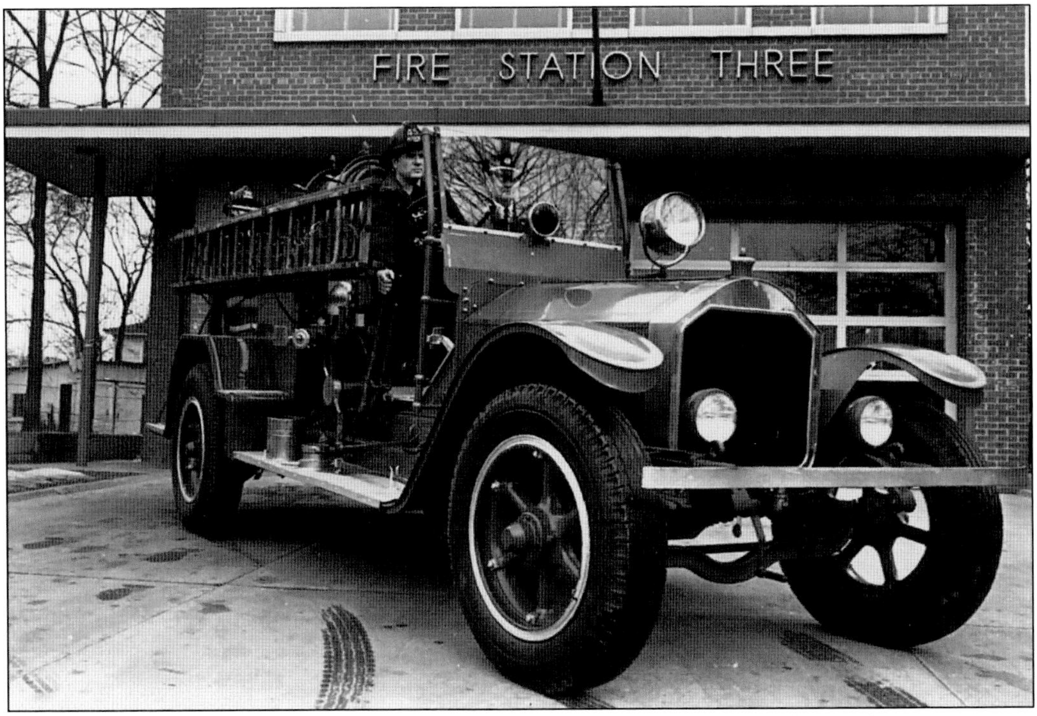

Reserve Engine, 1961. This 1926 American LaFrance Type 75 "triple combination pump, chemical, and hose motor car" remained a reserve unit until the early 1960s. It also served as a parade piece and was restored in 1987 by Col. Bob Biggs and members of the Raleigh Fire Department. Charlie Chappell is pictured in the driver's seat. (*RT* photograph by Bob Latham.)

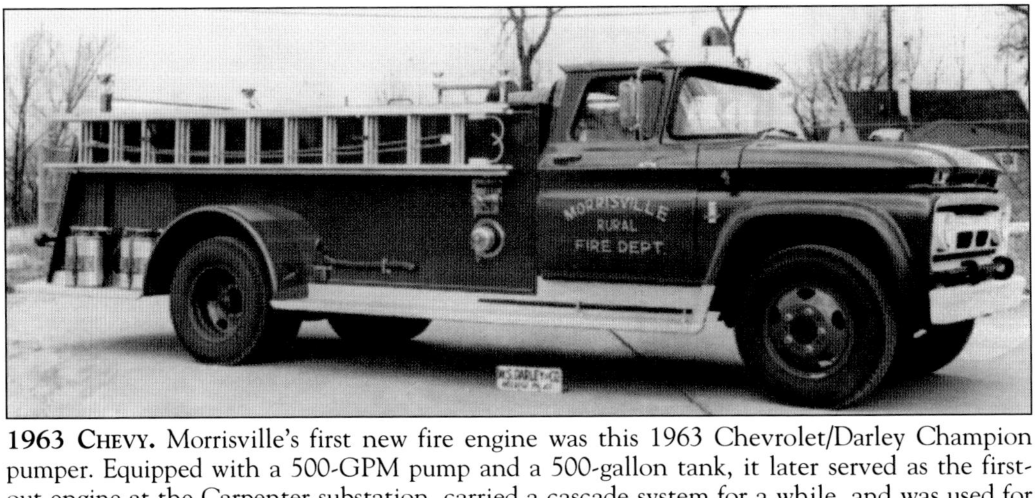

1963 Chevy. Morrisville's first new fire engine was this 1963 Chevrolet/Darley Champion pumper. Equipped with a 500-GPM pump and a 500-gallon tank, it later served as the first-out engine at the Carpenter substation, carried a cascade system for a while, and was used for some years at the Wake County Fire Training Center. Other Darley deliveries that decade included a 1964 Chevrolet pumper for Knightdale and a 1965 Chevrolet pumper for Durham Highway. (MRFC.)

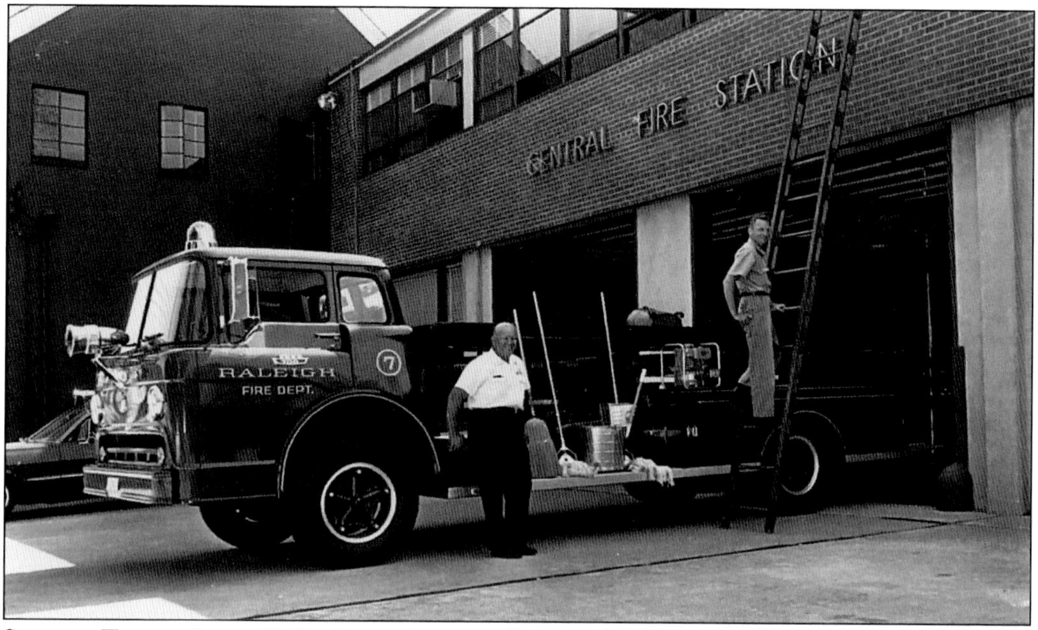

SERVICE TRUCK, JUNE 1963. Royce Lassiter (left) and Bill Bartholomew pose with Truck 7, the Raleigh Fire Department's newest ladder truck. Built on a new Ford chassis, firefighters constructed the rest of the truck themselves including using the ladder mounts from a retired 1922 service truck. The total cost was $5,000, compared to a factory-built price of around $35,000. (*RT* photograph.)

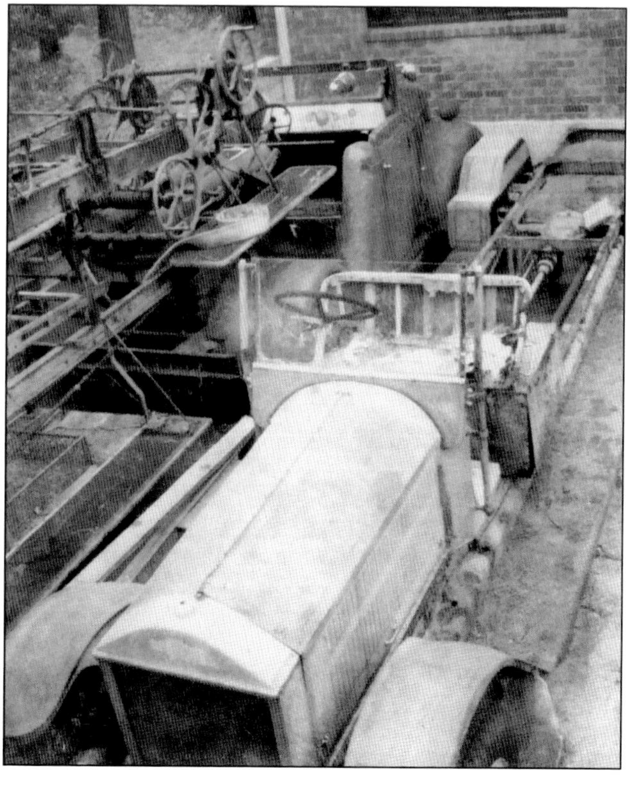

OLD TIMERS. With its body removed, the rest of Raleigh's 1922 American LaFrance Type 14 service truck (right) sits behind Station 8 on Western Boulevard. Originally delivered as a "combination service truck" with a chemical tank and hose reel, the motor was replaced by American LaFrance in 1929. It also received new wheels, axles, radiator, and cowling at the time. The ladder truck remained in service until 1962. Also pictured are the rusting remains of the city's 1939/1916 American LaFrance aerial ladder. (RFD.)

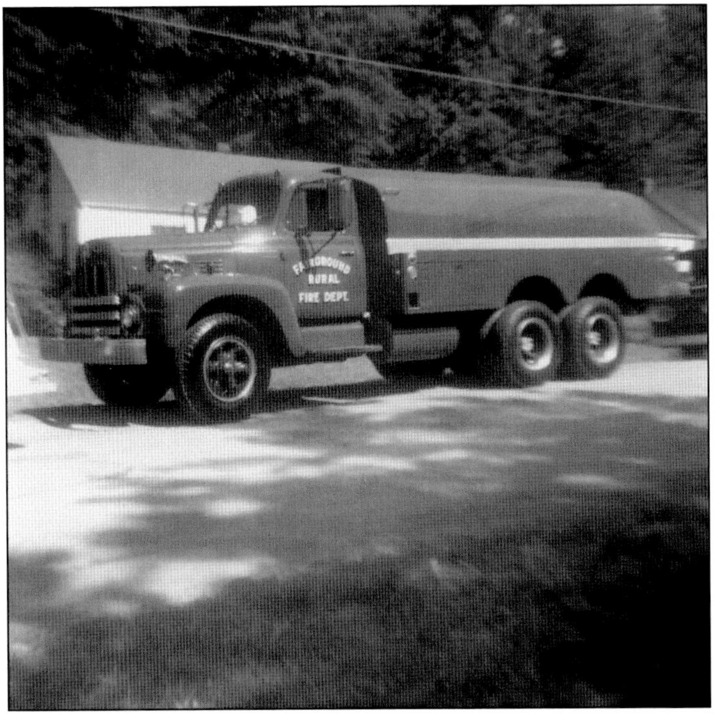

FAIRGROUNDS TANKER. This 1959 International tanker hauled 2,700 gallons of water for the Fairgrounds Fire Department. Converted fuel tankers, common in the early years, were difficult and sometimes dangerous to drive. Many didn't have baffles to prevent their contents from shifting; others had baffles, but were designed for lighter liquid contents than water. The extra tonnage also placed an added strain on a vehicle's suspension, brakes, and drive train. (WWFD.)

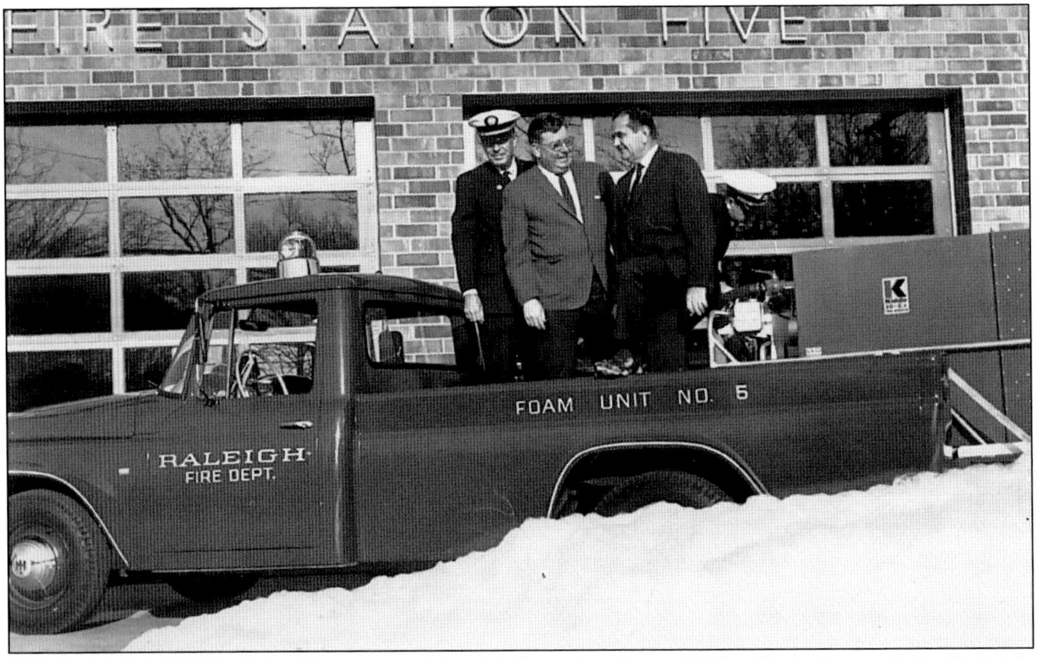

FOAM MACHINE, DECEMBER 1965. Cameron Village developer Willie York donated the Raleigh Fire Department's second foam machine, pictured here at Station 5. He was impressed after watching a demonstrator model at the Bryan Building fire in Cameron Village on December 2, 1964. Carried on a fire department-supplied truck, the $2,100 unit spit suds at a rate of 35,000 square feet a minute. Shown from left to right are Fire Chief Jack Keeter, York, Mayor Travis Tomlinson, and Capt. Rufus E. Keith. (*N&O* photograph.)

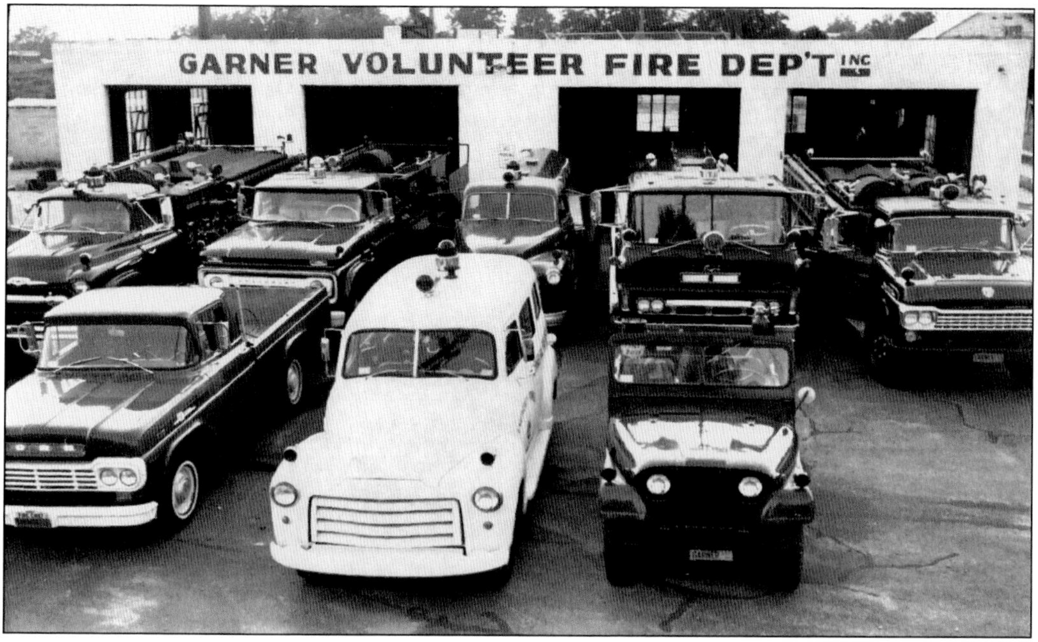

Garner Fire Department, c. 1966. Pictured at 110 Pearl Street are, from left to right, (front row) a 1958 Ford pick-up truck, a 1952 GMC panel truck, and a Jeep; (back row) a 1957 Chevrolet/American LaFrance pumper, a 1963 Chevrolet pumper, a 1948 Chevrolet tanker, a 1961 Chevrolet tanker, and a 1958 Ford/American LaFrance pumper. (GFD.)

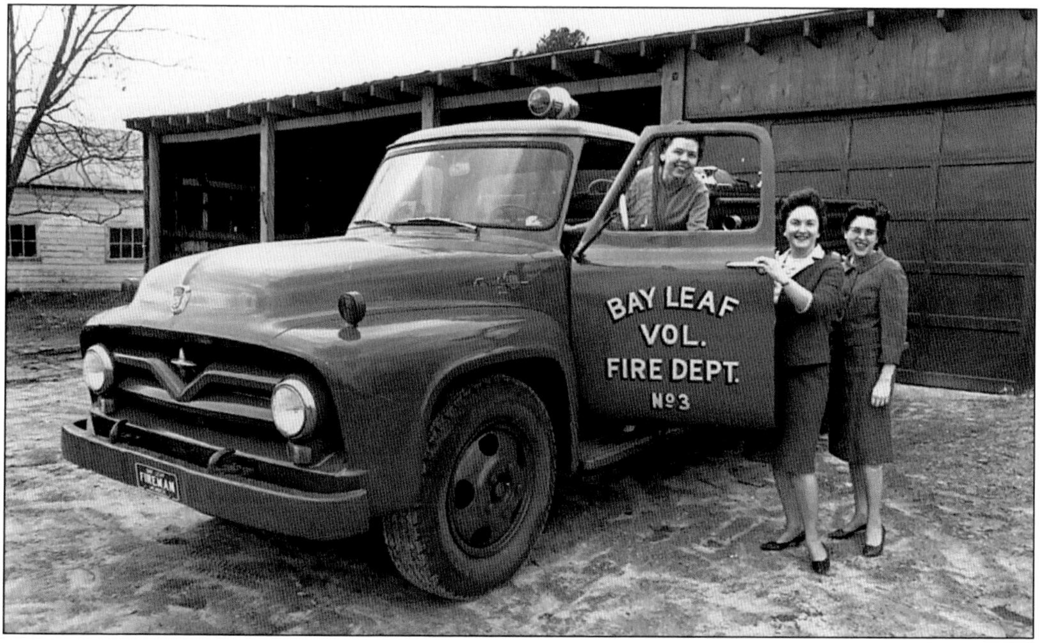

Widows of Living Firemen, February 1966. Posing with the $15,000 fire engine that their WOLF club helped purchase are, from left to right, the wives of Bay Leaf volunteer firefighters Bob Cooper, Ed Grice, and Bob Lassiter. The 1955 Ford was transformed from an Army weapons carrier into a 500-GPM pumper by C.W. Williams Company in Rocky Mount, North Carolina. (*N&O* photograph.)

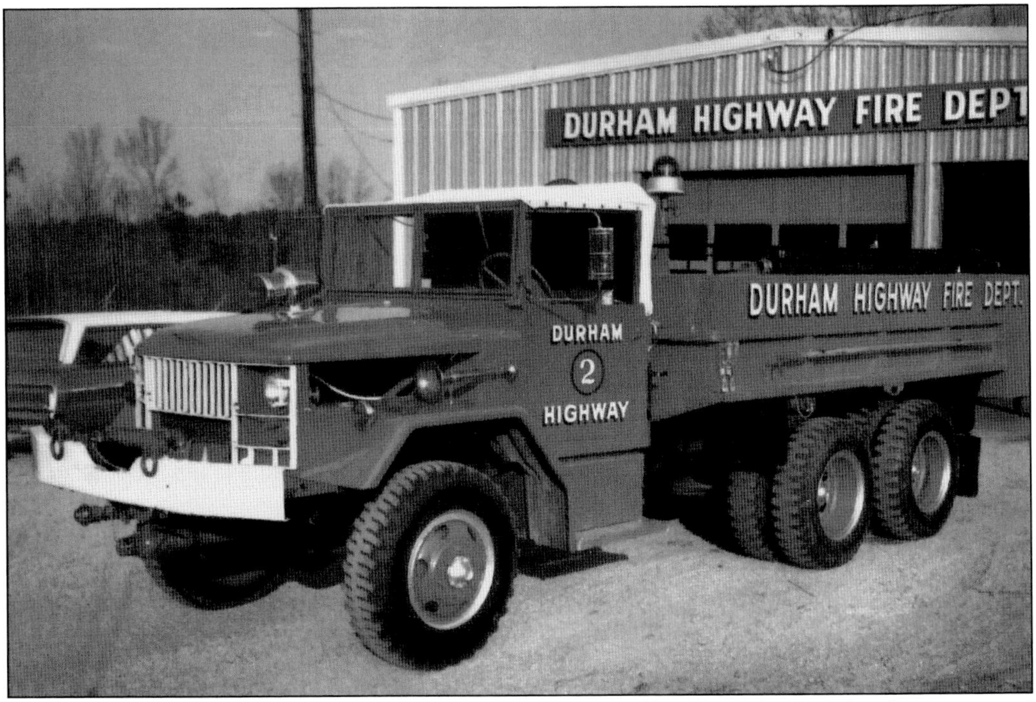

DEUCE-AND-A-HALF. This military surplus two-and-a-half-ton truck was the first piece of motorized equipment for the Durham Highway Fire Department. Acquired while waiting for their new 1965 Chevrolet/Darley pumper to arrive, firefighters converted the "deuce-and-a-half" into a fire engine for about $300. The 1,500-gallon unit was retired in 1969 and soon saw service with the new Falls Fire Department. (DHFD.)

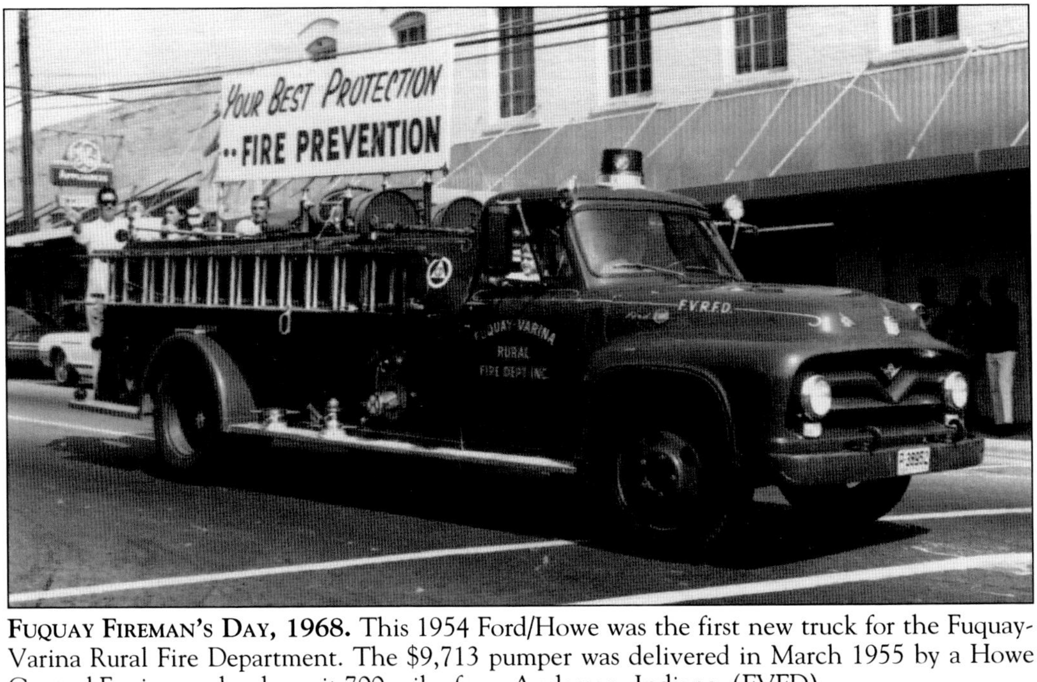

FUQUAY FIREMAN'S DAY, 1968. This 1954 Ford/Howe was the first new truck for the Fuquay-Varina Rural Fire Department. The $9,713 pumper was delivered in March 1955 by a Howe Control Engineer who drove it 700 miles from Anderson, Indiana. (FVFD)

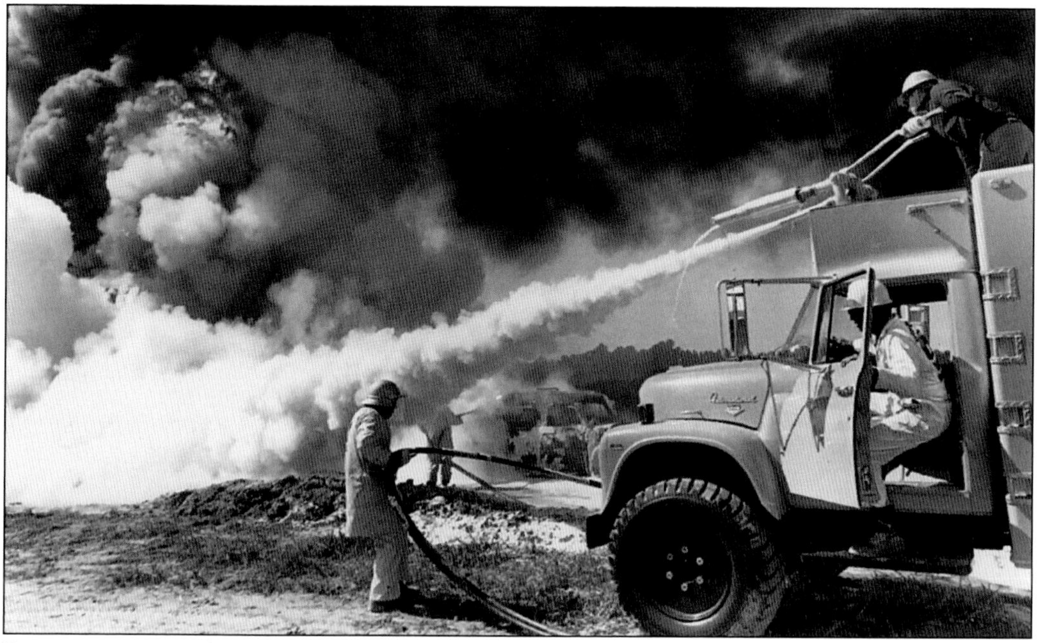

DEMONSTRATION, SEPTEMBER 23, 1969. Nine hundred gallons of jet fuel and 100 gallons of gasoline were ignited to demonstrate the capabilities of Raleigh-Durham Airport's newest crash truck. The 1969 International/Ansul Magnum 480 cost $32,000 and carried 1,350 pounds of Purple K dry chemical and 250 gallons of "light water." It was operated by ramp service personnel who were cross-trained in fire control. (*RT* photograph.)

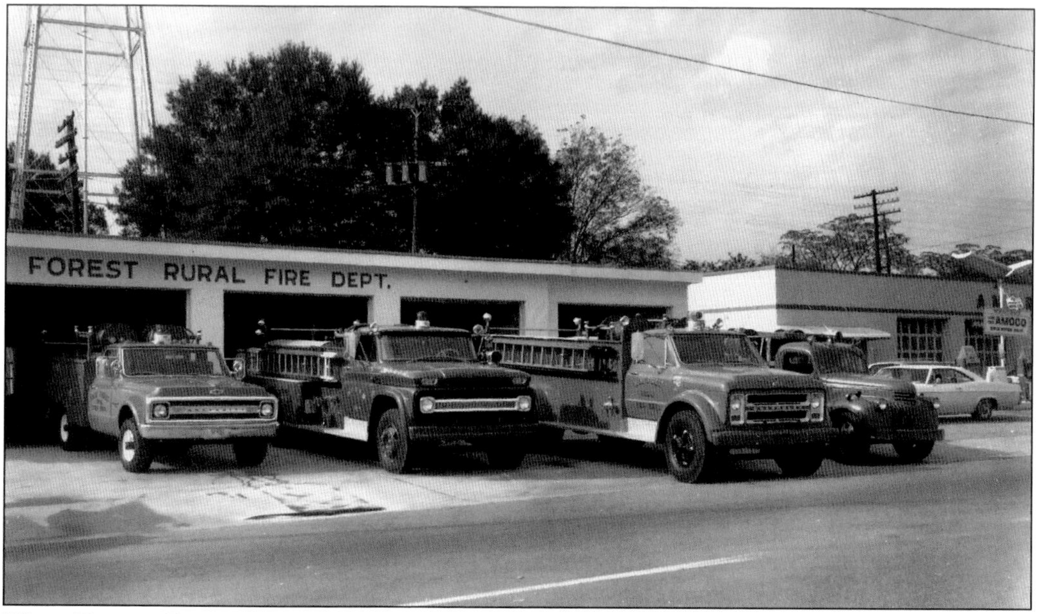

WAKE FOREST RURAL FIRE DEPARTMENT, 1969. Pictured in front of the fire station at 352 S. White Street are, from left to right, a 1969 Chevrolet/FWD/C.W. Williams brush truck, a 1963 Chevrolet/American LaFrance pumper, a 1968 Chevrolet/C.W. Williams tanker, and a 1942 Chevrolet pumper, their first fire truck. The rural fire department was formed in 1956; the fire station was completed in 1966. (WFFD.)

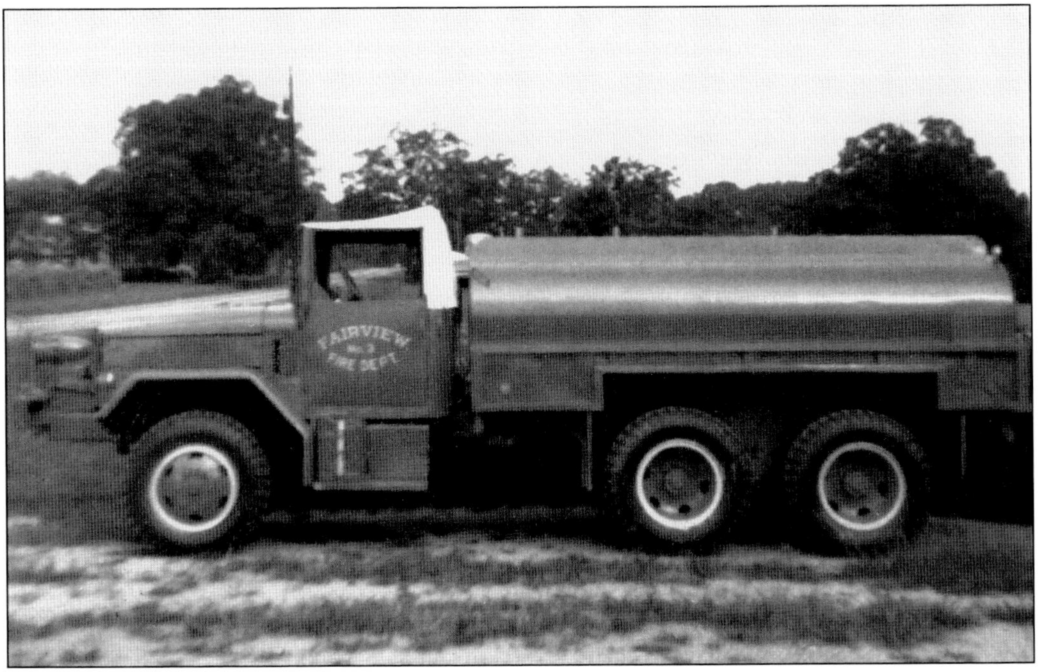

FAIRVIEW FIRE APPARATUS. "Big Red" was the first tanker of the Fairview Fire Department. The 1947 Diamond Reo, pictured above, was placed in service in 1965. The 10-wheeler converted gas tanker carried 1,000 gallons of water and had a small 80-GPM pump in the rear. Later modifications included a 100-GPM replacement pump that cost more than the truck itself. Big Red's replacement was a 1972 International/Atlas, pictured below left, that members assembled themselves. The 1,600-gallon tanker was placed in service in 1972 and is still used today. The fire department's first new truck was a 1968 International/American LaFrance, pictured below center. Delivered in 1969, the $17,764 fire engine was shipped by rail. The 750-GPM, 750-gallon pumper is still in service as a reserve unit. Members later built a 1972 Chevrolet brush truck, pictured below right, using a utility truck purchased from the state. It was placed in service in 1975 with a 350-GPM pump, 450-gallon tank, and two electric hose reels, all installed by the Equipment Committee. During the fire department's early years, committee members built and maintained all equipment. Duties included mounting, piping, wiring, and painting. (Courtesy of Fairview Fire Department.)

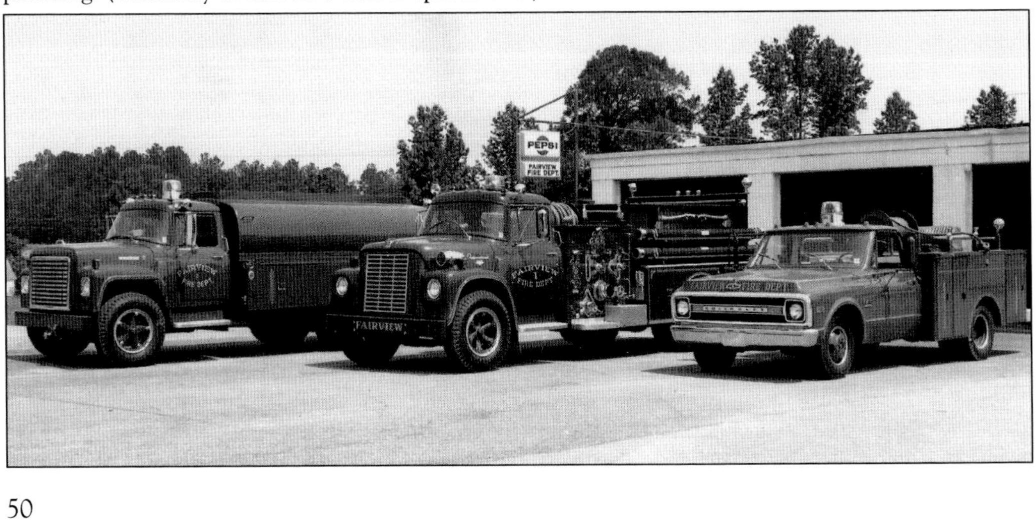

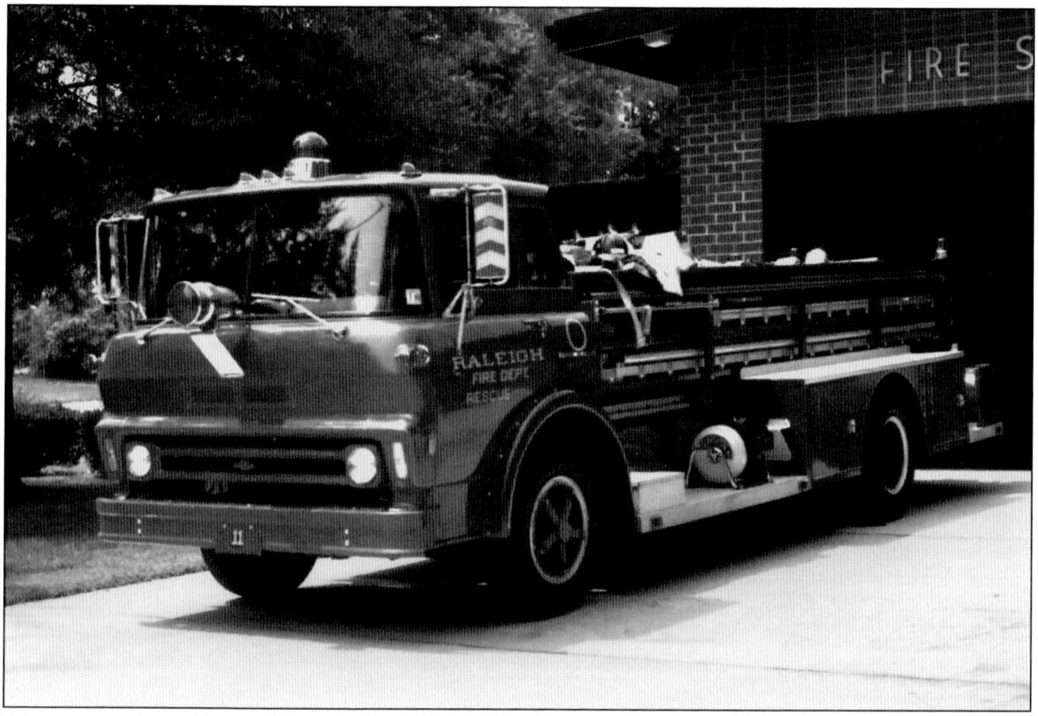

TRUCK 11. This 1971 Chevrolet ladder truck was placed in service after Raleigh Fire Station 11 opened at 2925 Glenridge Road. It joined two similar service trucks at Station 6 and Station 7—a 1963 GMC and 1963 Ford respectively. Replaced in 1986 by a 100-foot Seagrave aerial ladder, it later saw service as Truck 15. (Wayne Greer Jr. photograph.)

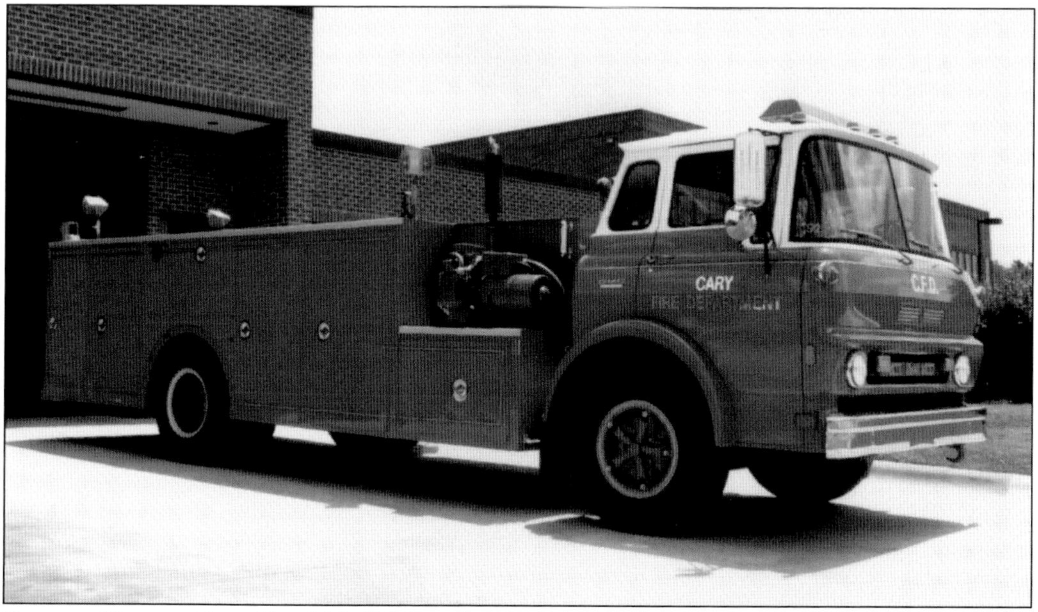

SERVICE TRUCK. Fire Chief Ned Perry designed the Cary Fire Department's second service truck, this 1975 GMC. The fully enclosed ladder truck proved so popular that its replacement, a 1995 Mack, was designed with a similar body style. (Lee Wilson photograph.)

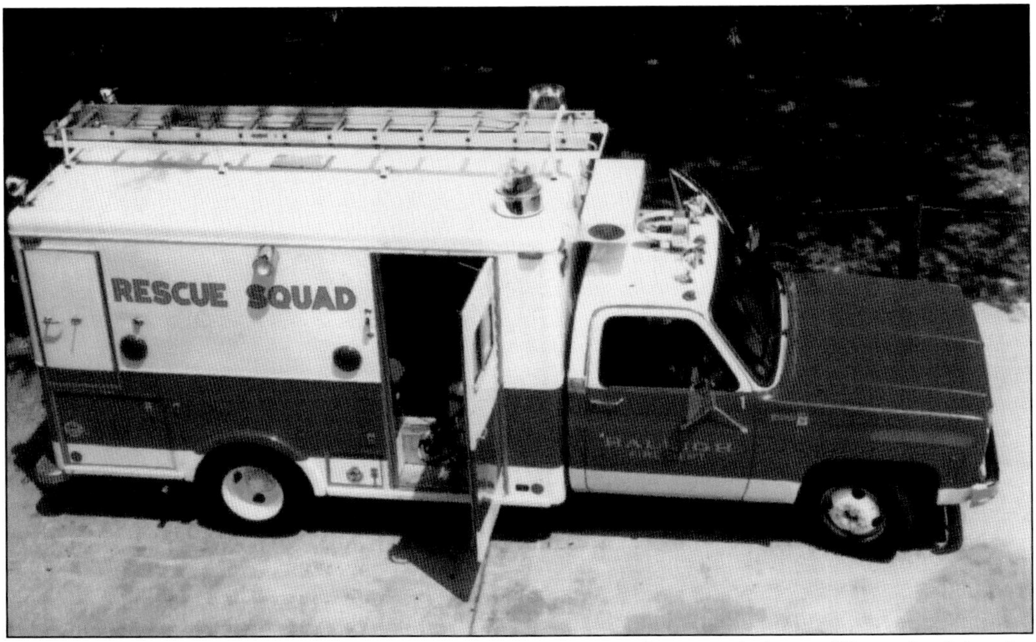

RESCUE SQUAD. The Raleigh Fire Department received two of these Chevrolet/Murphy ambulances in 1974–1975. Rescue 9 was placed in service with a 1974 model; Rescue 1 was assigned a 1975 model. (RFD.)

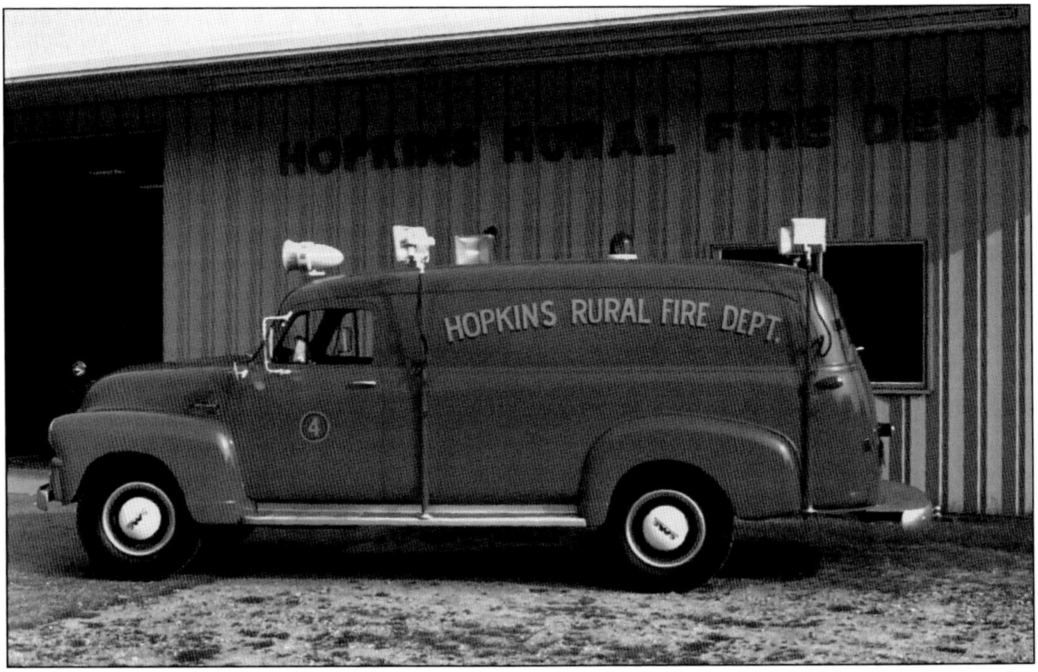

UNIT 4. Raleigh's original rescue truck was this 1954 GMC panel van that later served as the first vehicle of the Zebulon Rescue Squad. In 1978, it became a utility truck for the newly formed Hopkins Fire Department. (HFD.)

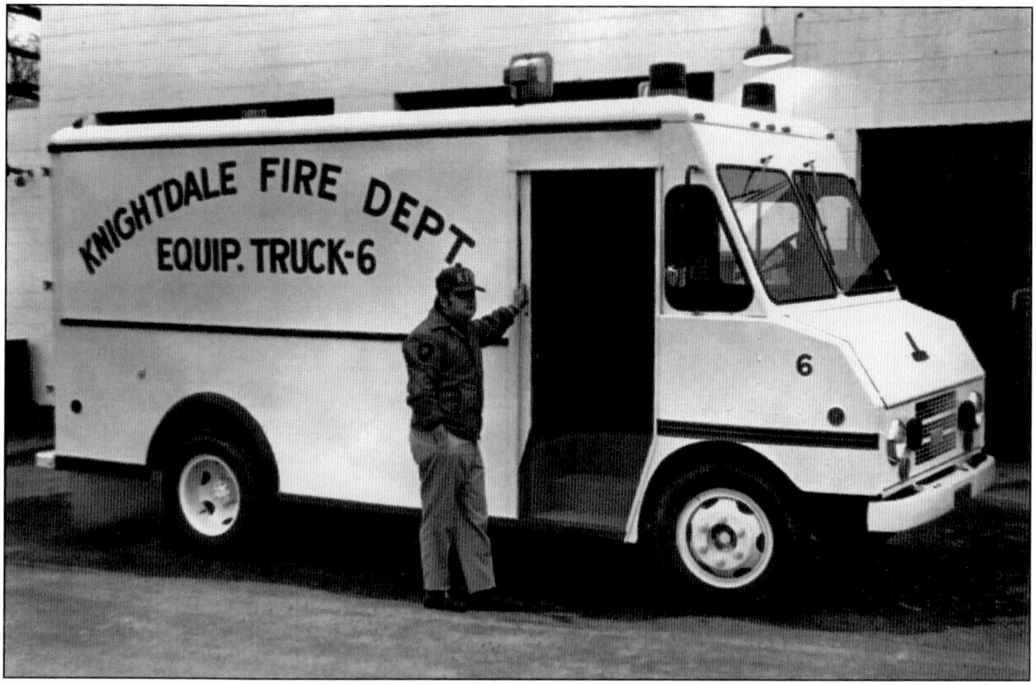

Equipment Truck, January 1977. Deputy Fire Chief George Gupton is pictured with the Knightdale fire department's newest addition, a 1974 Ford van equipped with floodlights, a medical oxygen generator, first aid supplies, blankets, stretchers, a generator, two hand lights, shovels, mops, brooms, reels of electric cord, a smoke fan, two air packs, an axe, and other leverage tools. (*GLF* photograph.)

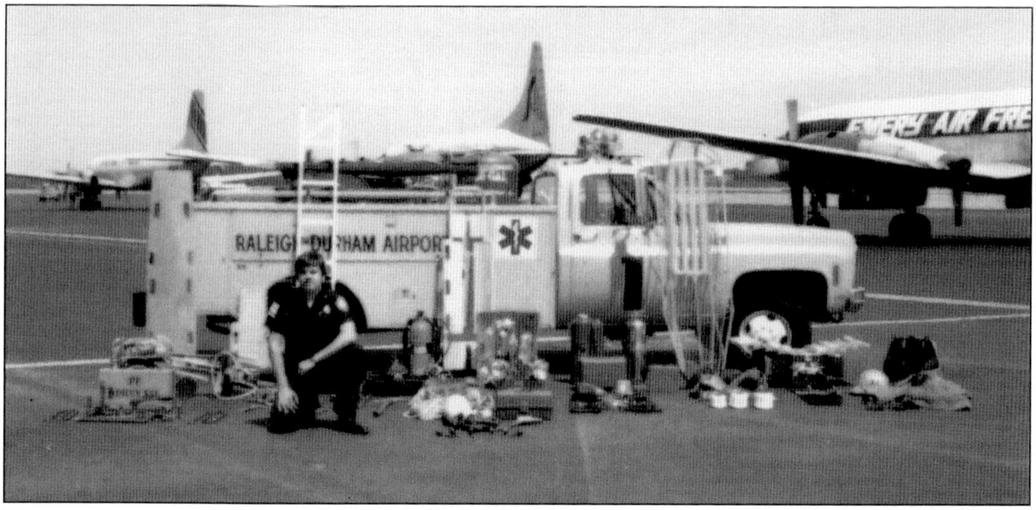

Rescue Equipment, 1980. Airport firefighter Dale Johnson poses in front of CFR-1, a 1979 Chevrolet/Ansul/Reading rescue and dry-chemical unit. Displayed equipment includes a Hurst power tool spreader and power unit; long and short back boards; 10-foot folding attic ladder; hand tools including a hacksaw, crash axe, and sledgehammer; air chisel kit; ropes; SCBA cylinder for air chisel; scoop stretcher; triage tag packs; MAST trousers; fire extinguishers; traction and air splints; Stokes basket with straps and blankets; cervical collars; medical/trauma box; and turnout gear. (Courtesy of Dale Johnson.)

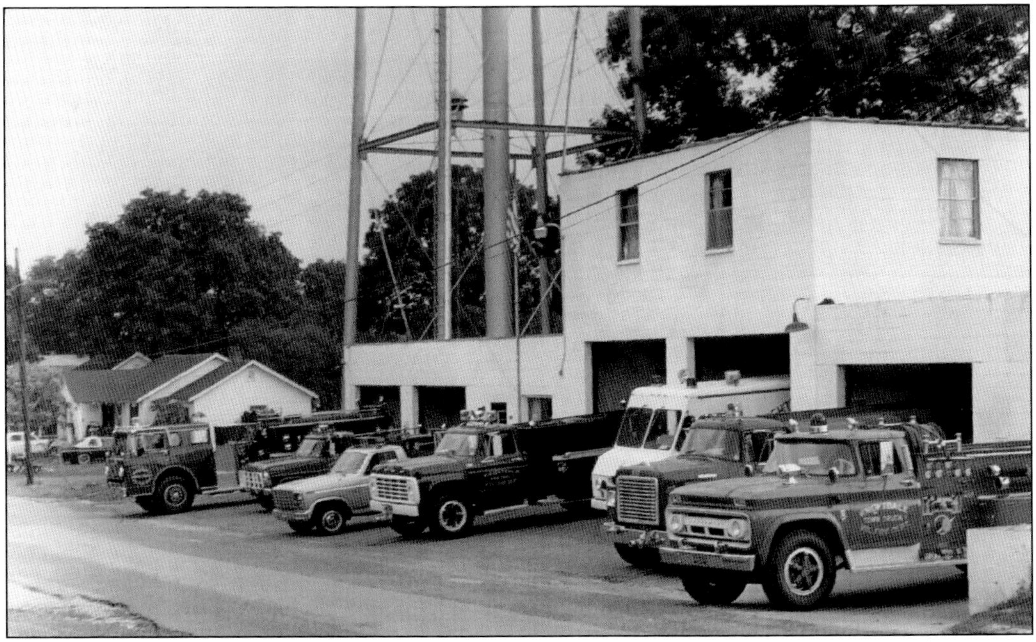

KNIGHTDALE FIRE DEPARTMENT, C. 1982. Pictured from left to right in front of the Hester Street fire station are a 1978 Ford/Howe pumper, a 1970 Ford brush truck, a 1980s Ford pick-up truck, a 1973 Ford/Atlas tanker, a 1974 Ford equipment van, a 1976 International tanker, and a 1963 Chevrolet/Darley pumper. (EWFD.)

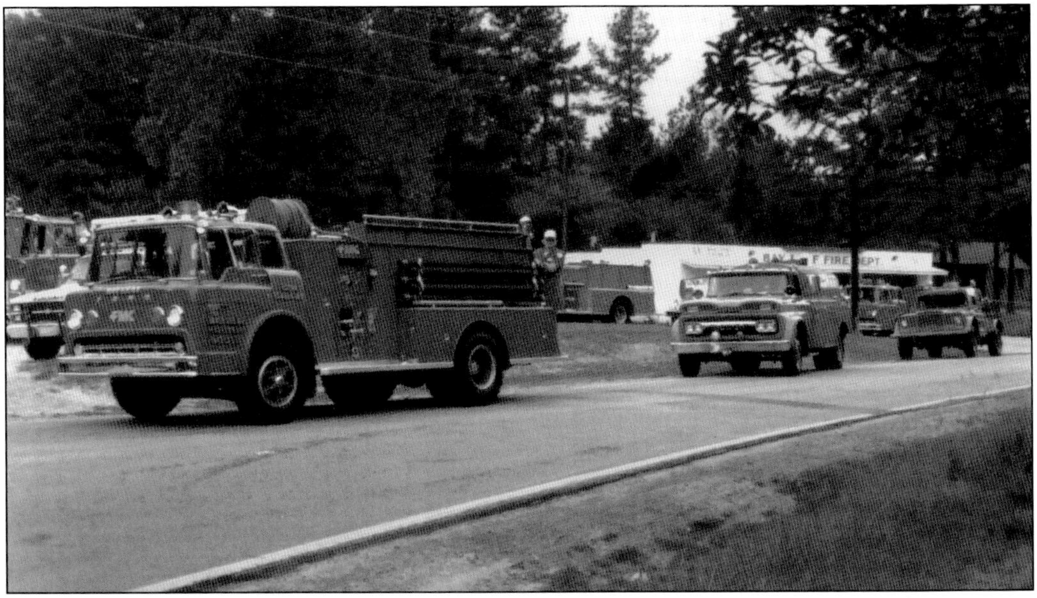

MOVING DAY, OCTOBER 9, 1983. Bay Leaf fire apparatus is pictured in ceremonial procession from their old station at 11617 Six Forks Road to their new quarters at 11713 Six Forks Road. Opening day ceremonies included a ribbon cutting, recognition of guests, and refreshments served by the volunteer firefighters' wives. The apparatus are, from left to right, a 1979 Ford/FMC pumper-tanker, a 1962 GMC tanker, and a 1960s Kaiser brush truck. (BLFD.)

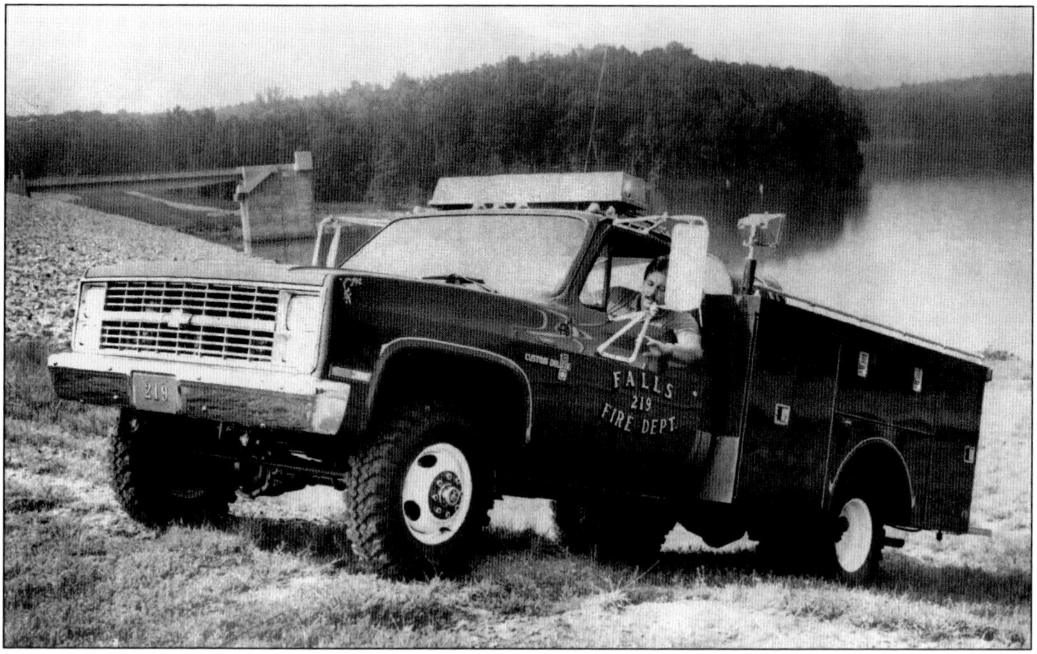

DAM FIRE TRUCK, C. 1984. Engine 213, a 1984 Chevrolet/Hale brush truck, climbs an access road at the Falls Lake Dam with Neal Blackmon at the wheel. The 18-square-mile reservoir was created by the Army Corps of Engineers three years earlier. (FFD.)

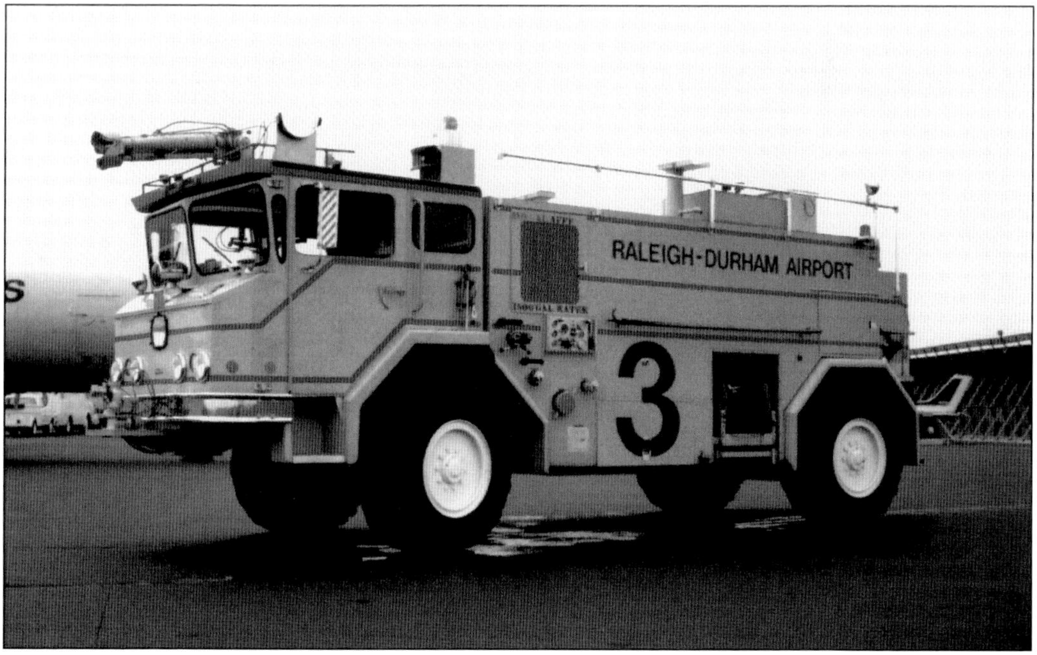

CRASH TRUCK, 1984. This Walter B1500 airport crash truck, purchased new in 1977, carried 1,500 gallons of water and 180 gallons of foam. Its 1000-GPM pump was powered by a separate diesel engine. The truck was designed so it could use 75 to 80% of its rated capacity with a one-man crew. (Dale Johnson photograph.)

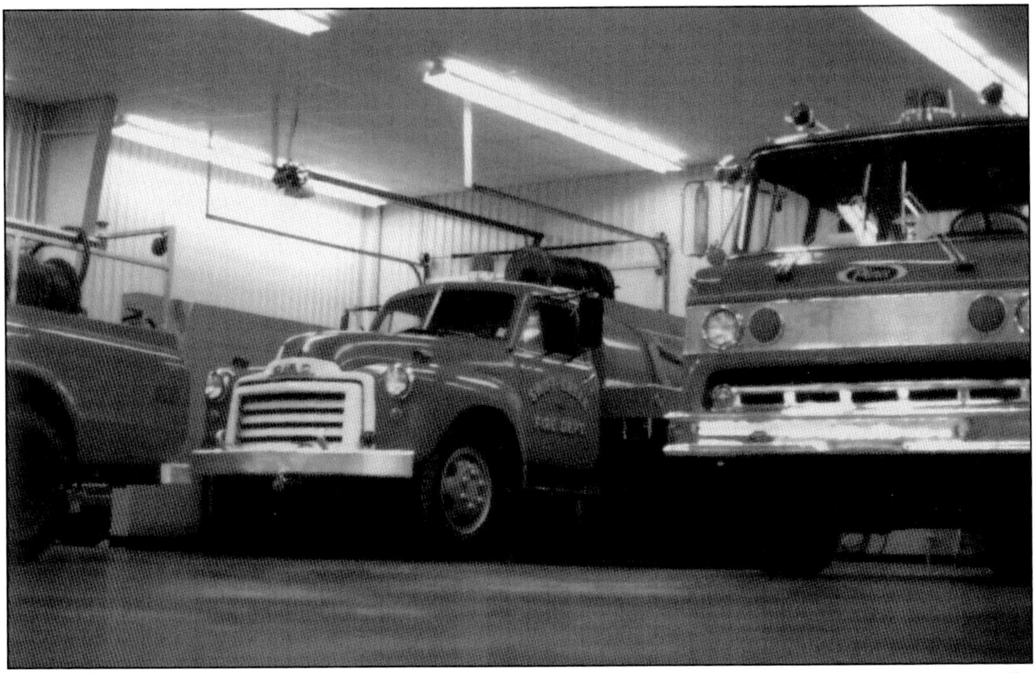

SWIFT CREEK FIRE DEPARTMENT, C. 1984. Pictured inside the apparatus bay at 6000 Holly Springs Road are a 1951 GMC tanker (left) and a 1976 Ford/Pierce pumper (right). The rear of a 1969 Chevrolet brush truck is visible on the far left. (SCFD.)

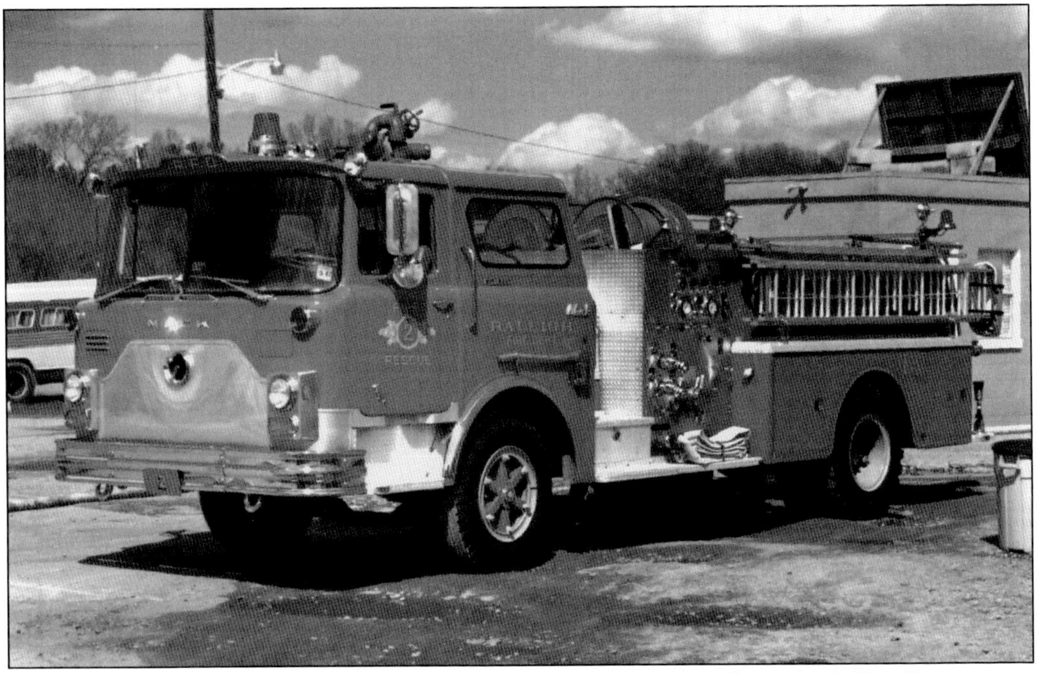

ENGINE 2. This 1975 Mack CF pumper was one of 16 delivered to the Raleigh Fire Department during the 1970s and 1980s. Model years ranged from 1970 to 1971. Most were later repainted white over red and received upgraded warning lights and Federal Q sirens. Other changes added cross lay compartments and replaced hard suction hoses with ladders. (Tommy Rich photograph.)

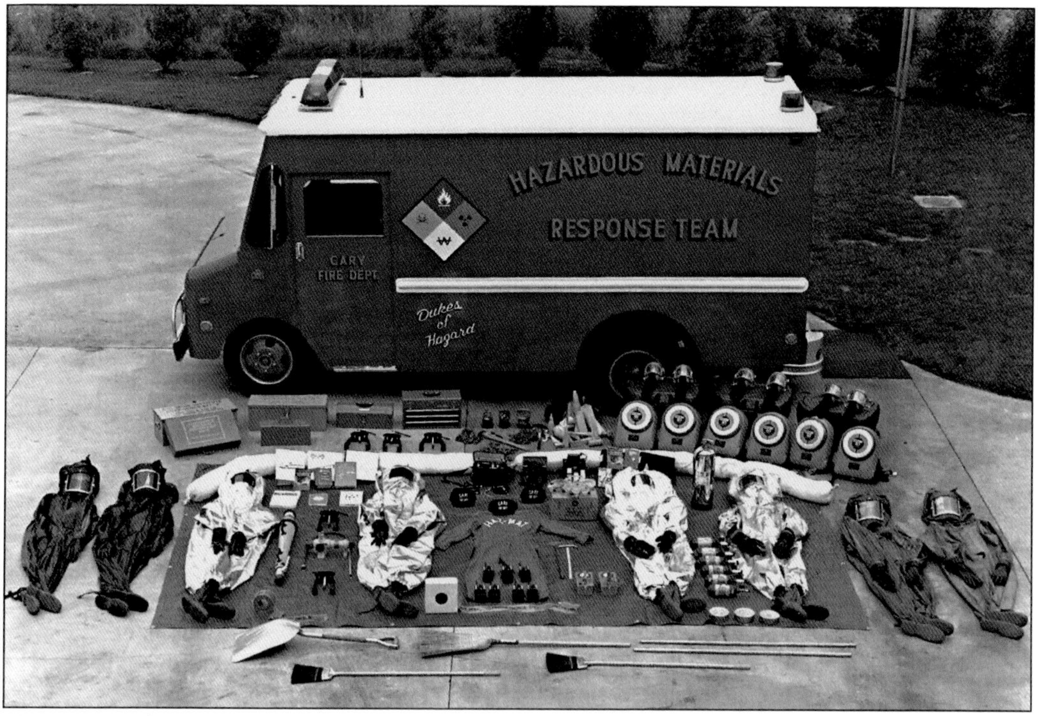

Haz-Mat. This 1986 Dodge step van transported the equipment of the Cary Fire Department's hazardous material team. Formed in 1986, the short-lived squad operated until about 1990. Raleigh and Wendell also formed haz-mat teams in the 1980s. Both are still active. (CFD.)

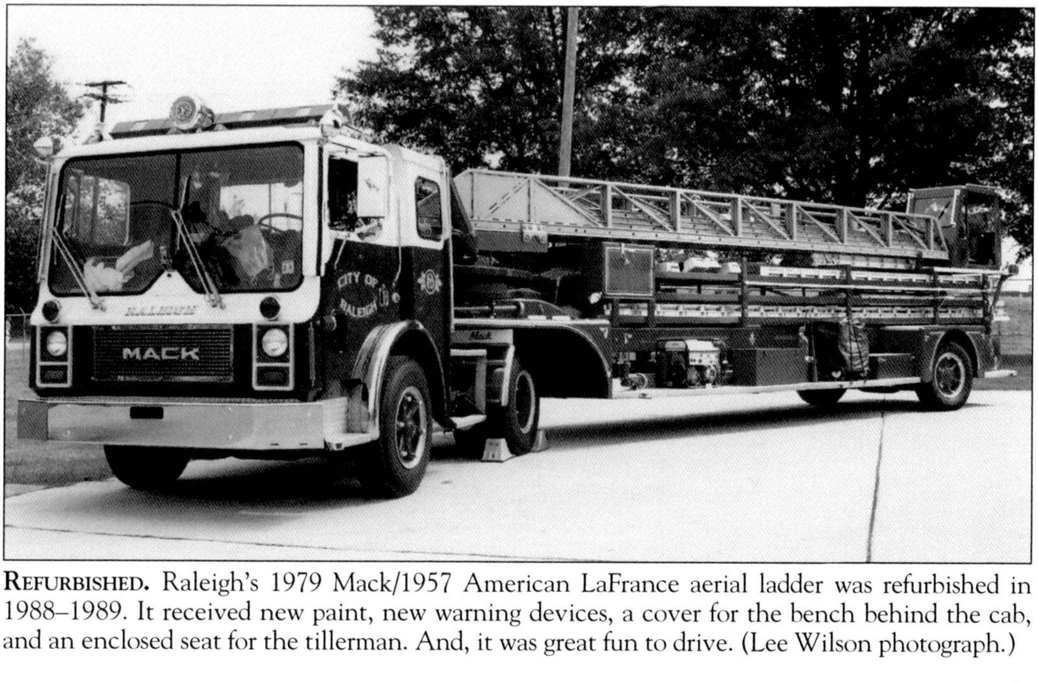

Refurbished. Raleigh's 1979 Mack/1957 American LaFrance aerial ladder was refurbished in 1988–1989. It received new paint, new warning devices, a cover for the bench behind the cab, and an enclosed seat for the tillerman. And, it was great fun to drive. (Lee Wilson photograph.)

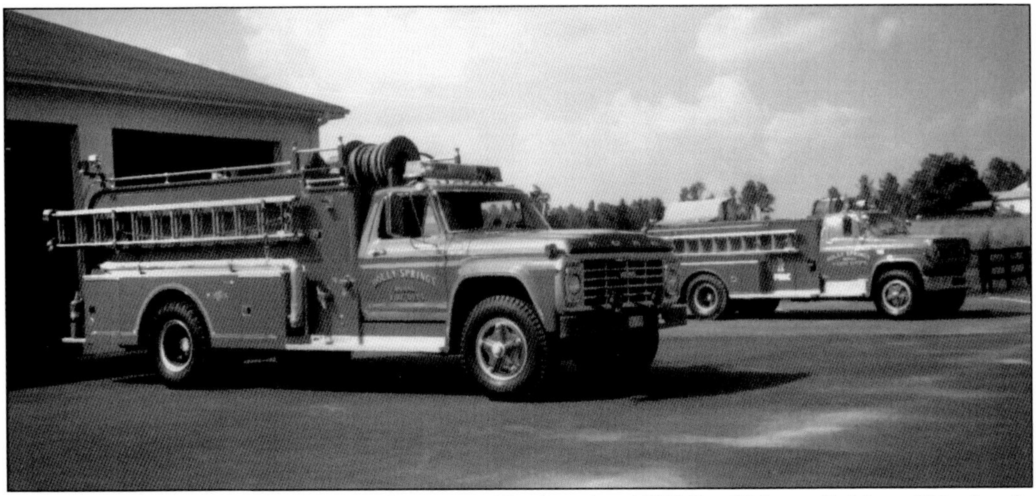

Holly Springs Fire Apparatus. Shown are Tanker 4, a 1977 Ford/Howe (left) and Tanker 3, a 1985–1986 Chevrolet/FMC. The former was purchased in 1990 and sold in 1992–1993. The latter was purchased new in 1985–1986 and removed from service in 2003. Other Holly Springs apparatus at the time included a 1974 Ford/Bean pumper and a 1986 Kenworth/FMC pumper-tanker. (Lee Wilson photograph.)

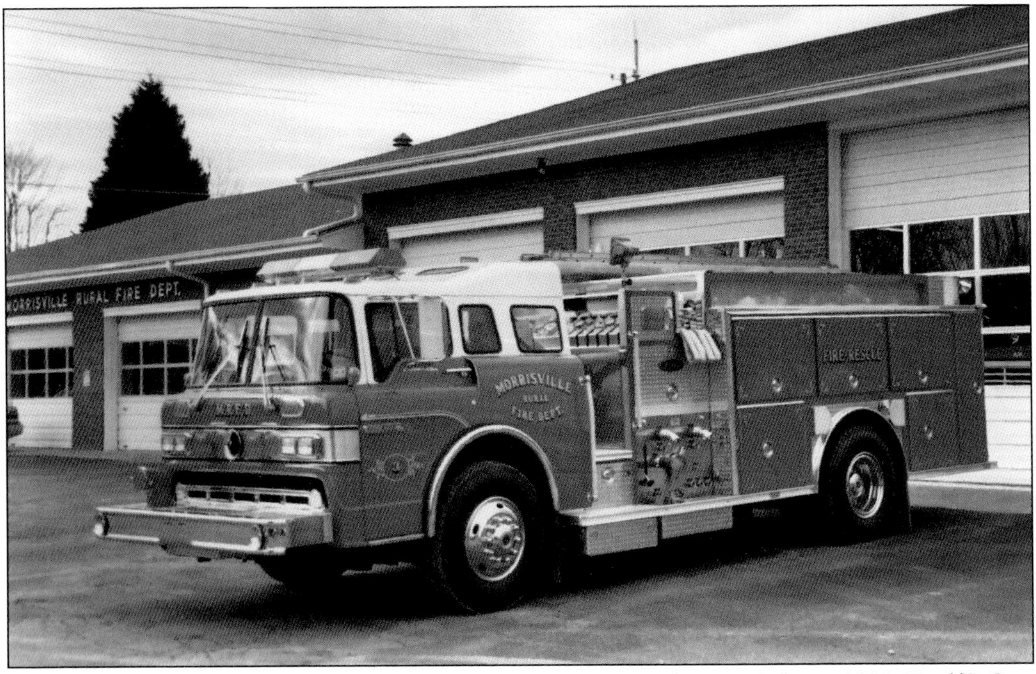

Pumper-Tanker. The Morrisville Fire Department received two of these 1989 Ford/E-One pumper-tankers at the end of the decade. This one is pictured at Station 1 at 100 Morrisville-Carpenter Road before the remodeling that repositioned the apparatus bays to face north instead of east. (Jeff Harkey photograph.)

Three

FACILITIES

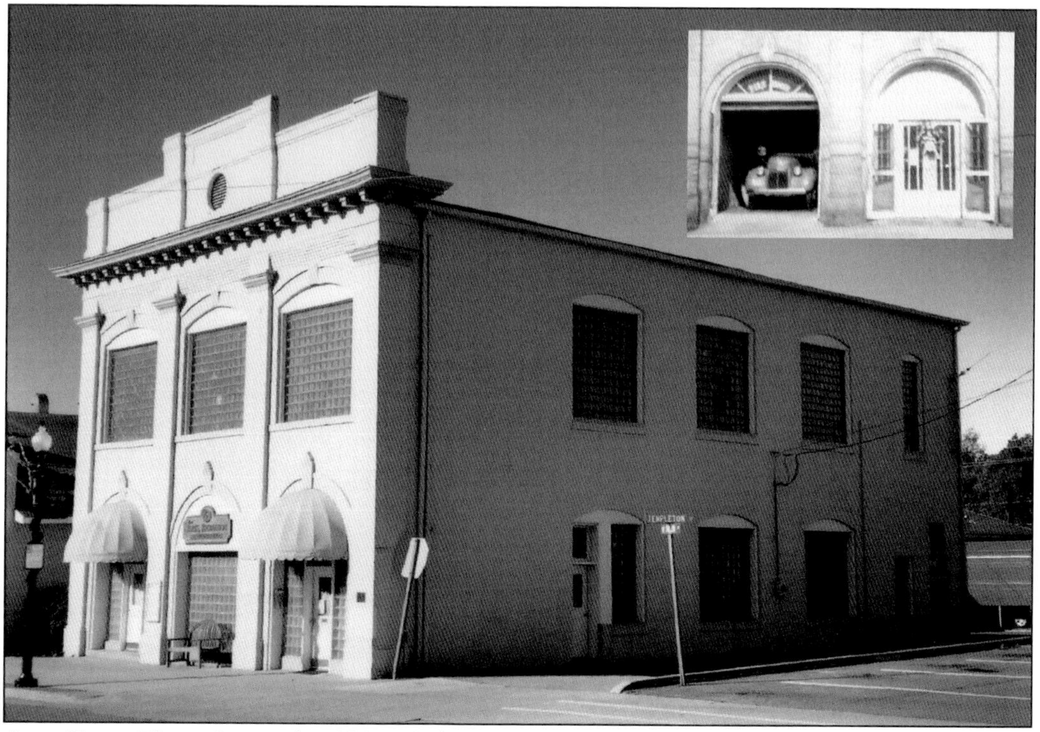

APEX TOWN HALL. Located at 237 N. Salem Street, the old town hall housed the fire department until 1956. The fire truck was stored in a single bay in the front of the 1912 building. When rural service was added in 1952, the additional fire trucks were stored in sheds around back. (Mike Legeros photograph, inset courtesy of Andy Clark.)

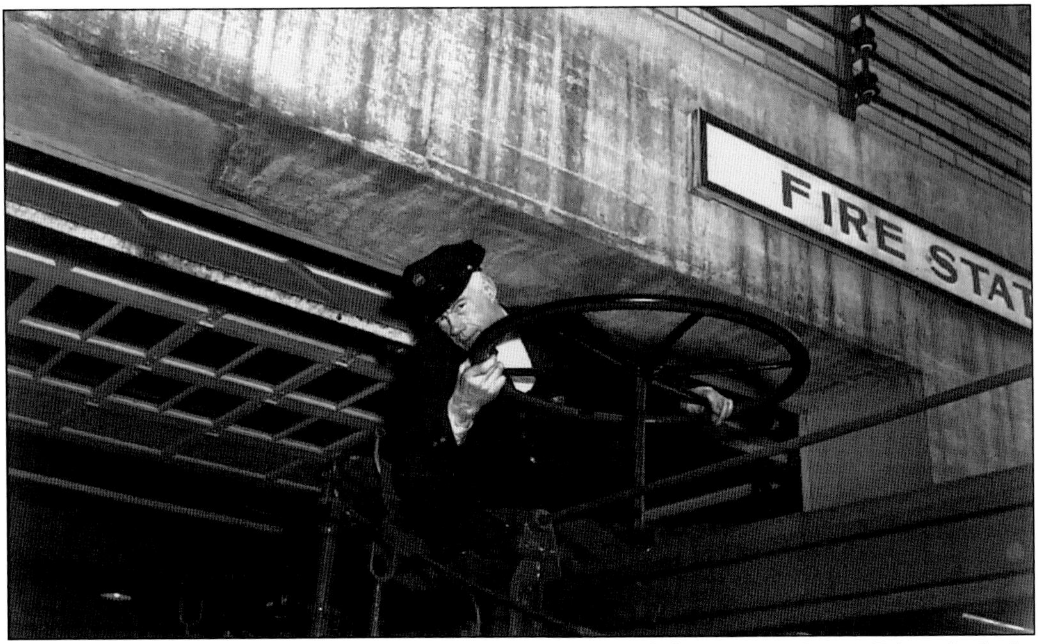

STATION 2, JANUARY 1950. Pictured above, Raleigh firefighter Boaz "Bozo" Edwards ducks while steering the rear of the old American LaFrance aerial ladder. Pictured below, sitting down for supper, clockwise starting on the left, are Edwards, Fire Chief Alvin Lloyd, Capt. Charlie Hayes, May "Ace" Parker, Capt. Jimmy Blake, Norman Conyers, Thomas Hall, Carl "Tootsie" Wall, and Willis Kirk. (*N&O* photographs.)

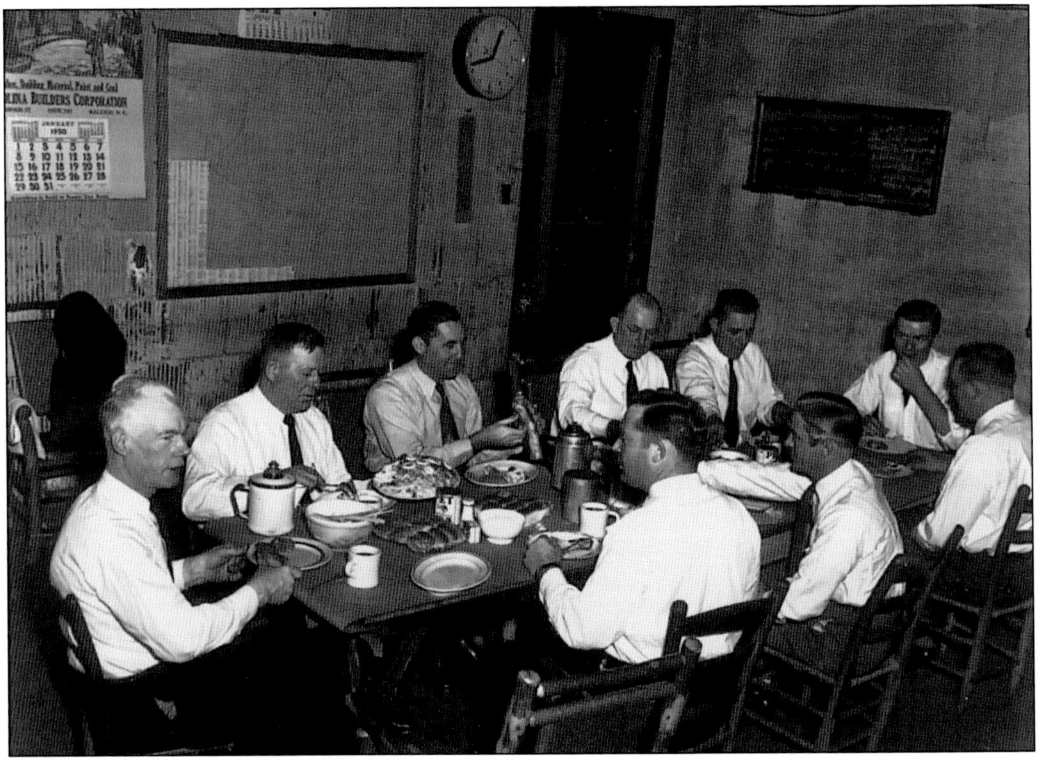

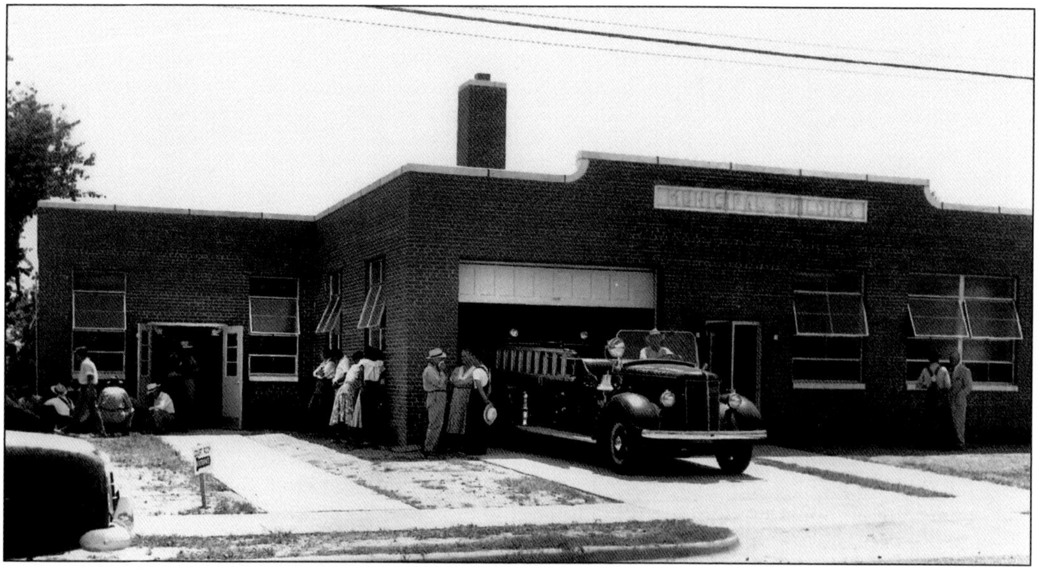

MUNICIPAL BUILDING, JULY 1951. Located at 131 Fuquay Avenue, this new building housed the Fuquay Springs fire, police, and water departments, a jail, a recorder's court, and offices for the town. Construction was started about a year ago but was halted for six months due to a shortage of steel. Pictured in front is a 1945 Mack Type 45 pumper. The truck was purchased in 1946 after a disastrous downtown fire prompted an emergency resolution by the Town Board to upgrade Fuquay's firefighting capabilities. A bid by the Mack Truck Company was presented and accepted for $5,555.80. (*N&O* photograph by Lawrence Wofford.)

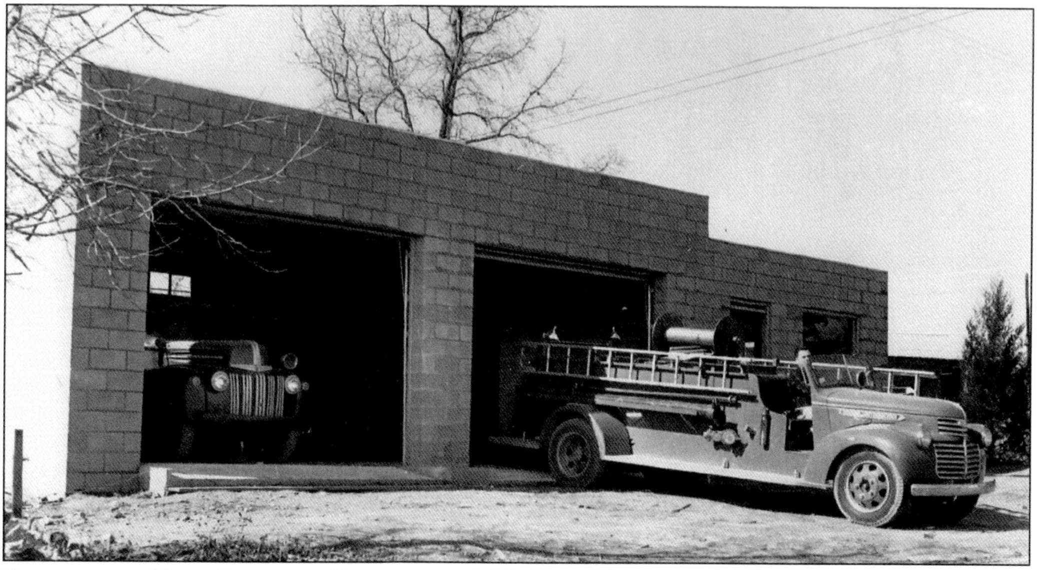

KNIGHTDALE FIRE STATION, FEBRUARY 1956. Fire Chief Randolph Griffin pulls the 1942 pumper out of the nearly finished fire station. For three years, fire equipment had been stored in the garage of Jim Keith. The concrete-block building on Hester Street also contained a large room to be rented to the town as office space. Later additions included two apparatus bays on the north side of the station in 1959 and an upstairs recreation room complete with pool table in 1967. (*N&O* photograph.)

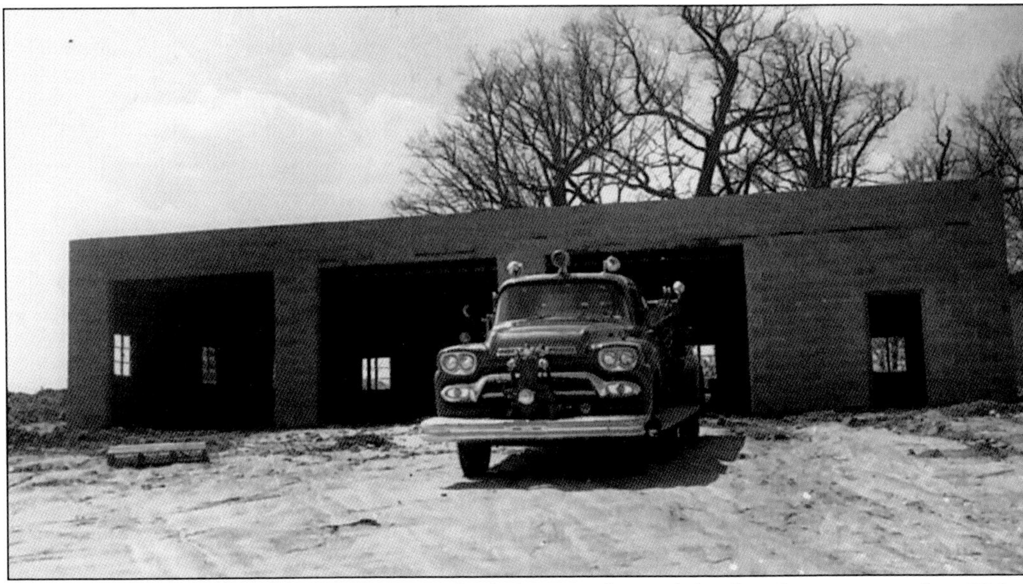

New Hope Fire Station, April 1958. Built by members of the fire department, this 60-by-40 foot cinderblock building on New Hope Road soon housed three pieces of equipment and office. Later additions included five more bays and expanded office space. (*RT* photograph by A.C. Snow.)

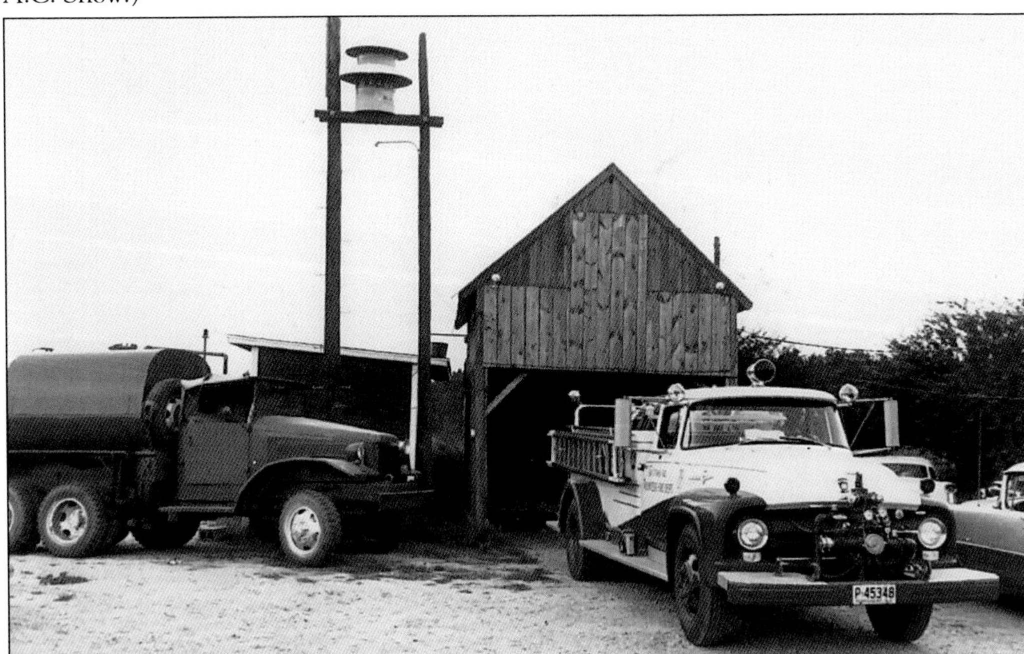

Six Forks Road Fire Department, September 1958. First housed in this donated barn beside Howell's Store on the west side of Six Forks Road just south of Lassiter Mill Road, the volunteer fire department soon moved a block north to a metal building on the east side of Six Forks Road. Also pictured are their first two pieces of apparatus, a 1,500-gallon military surplus tanker (left) and a 1956 Ford/American pumper with front-mounted Barton pump. (*N&O* photograph.)

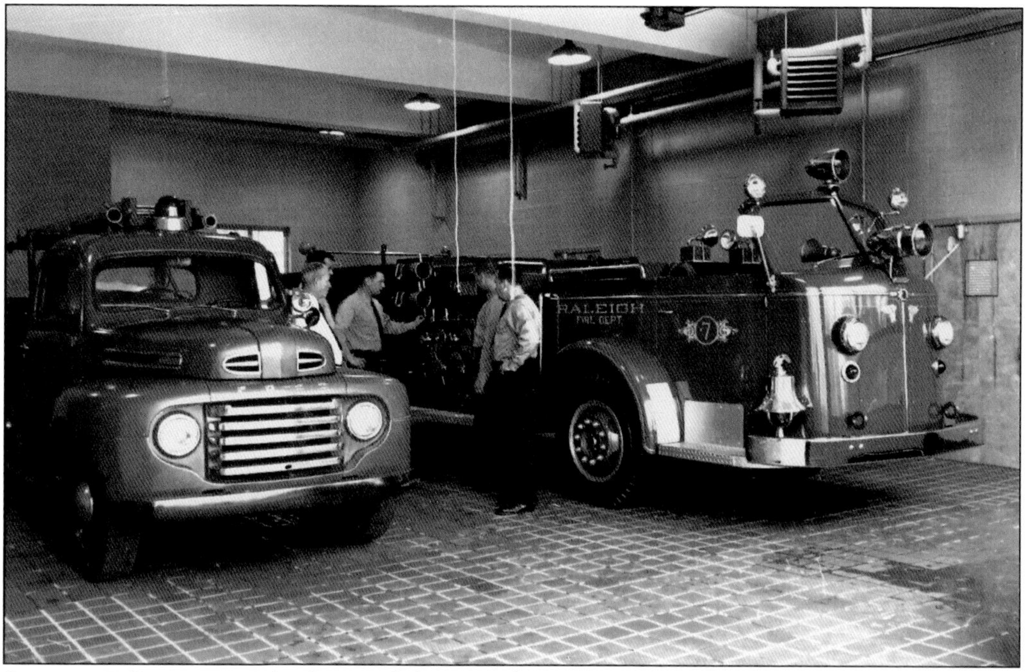

STATION 7, JANUARY 1960. Pictured above in the apparatus bay of Raleigh's newest fire station at 2100 Glascock Street are a 1948 Ford F3 auxiliary truck (left) and a 1953 American LaFrance 700 Series pumper (right). Pictured below watching television in the day room are, from left to right, Melvin Williams, Capt. Virgil Mims, Ben Dixon, Wilbur Beasley, and James Warren. Station 7 was the first single-story fire station built since Station 4 opened on Jefferson Street in 1924. The station also marked the debut of a basic design used on 18 subsequent new and replacement Raleigh fire stations. (*N&O* photographs.)

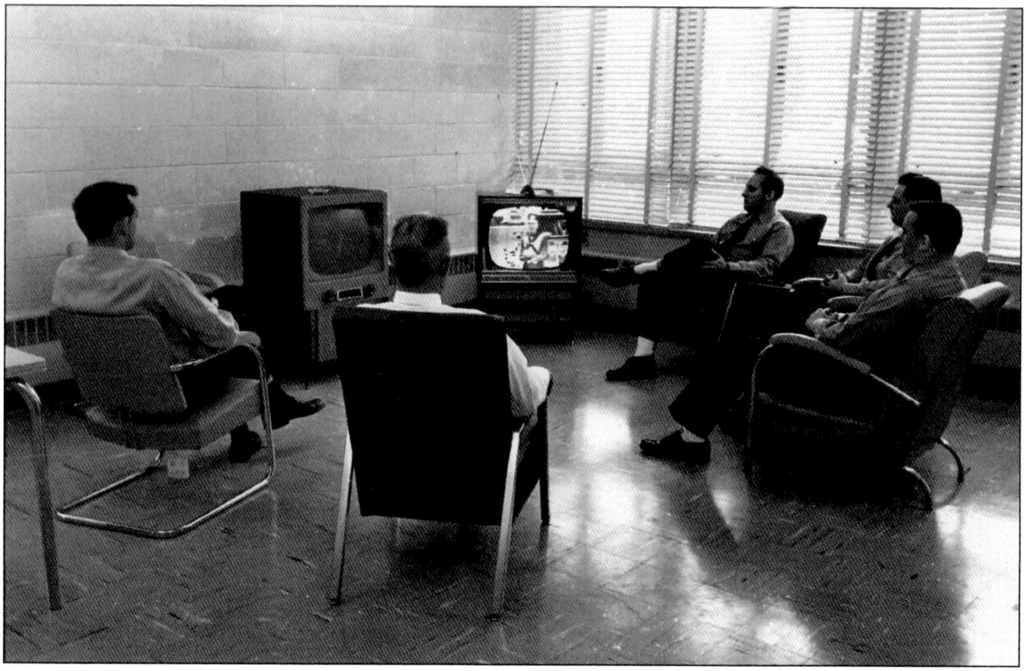

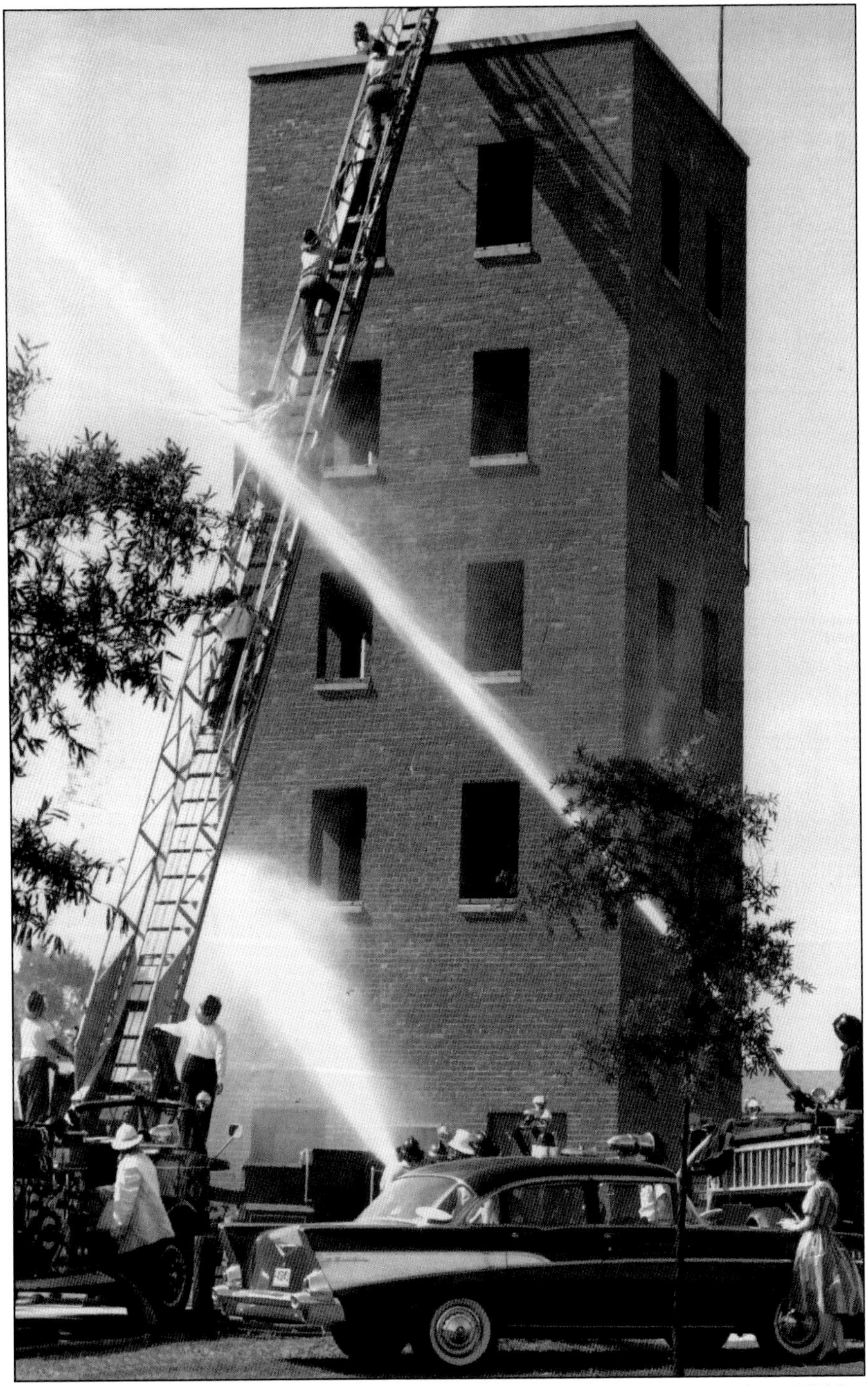

Drill Tower, May 1959. This five-story training tower was constructed in 1955 in south Raleigh. Though only 20 feet square, the structure was considered as high as an average building. The fire department's old training tower, located at old Station 1 on W. Morgan Street, was demolished along with the rest of that building in 1941. The new tower, plus the pavement around the base, cost about $30,000. (RFD.)

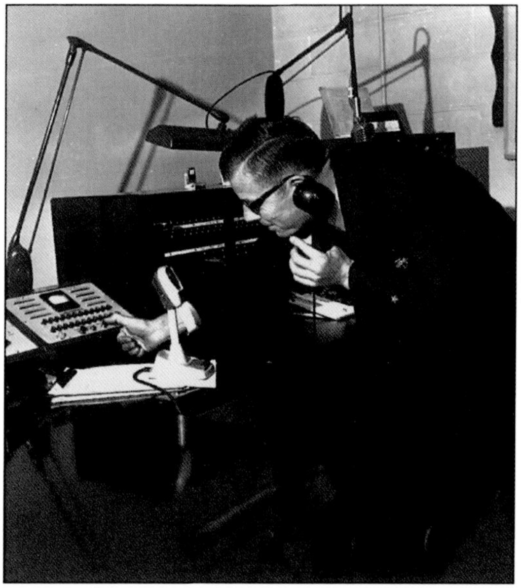

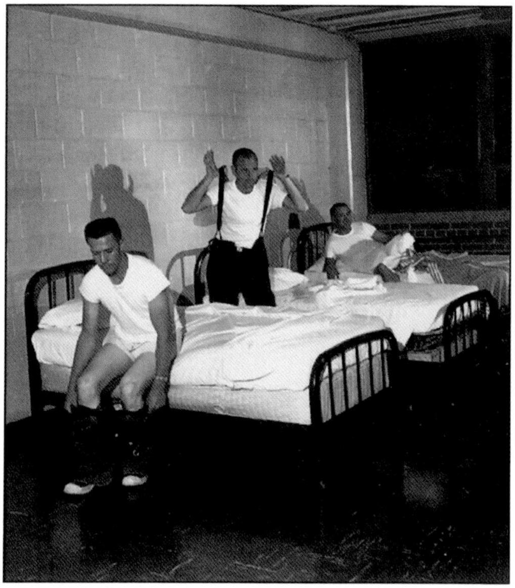

THE SEQUENCE, JUNE 1960. Raleigh Fire Station 1 is dark and quiet at 3:48 a.m. Upstairs, 20 firemen are asleep in the dormitory. Downstairs, dispatcher Roy High is smoking a cigarette and reading an outdated magazine. He walks to the door of the apparatus floor. In the dim light he sees the three American LaFrance fire engines: two pumpers and an aerial ladder. Their red paint looks orange. High lights another cigarette and resumes his reading. Half an hour later the phone rings. "Fire department," he answers before the first ring is finished. His hand is already reaching for the alarm button as he asks their address. Bright lights bring the station to life as a gong starts to sound. Upstairs, shown left to right, Charlie Narron, Rufus Keith, and Reginald "Tex" Barnette jump out of bed and into their "bunker pants." Down the poles they slide. In swift silence the firefighters take their positions. There are five men on each truck—the driver behind the wheel, the captain to his right, and three firemen on the tailboard. Doors are raised and within 18 seconds from the first ring of the telephone, Engine 1, Engine 9, and Truck 1 are rolling onto Dawson Street. (*RT* photographs by Bob Latham.)

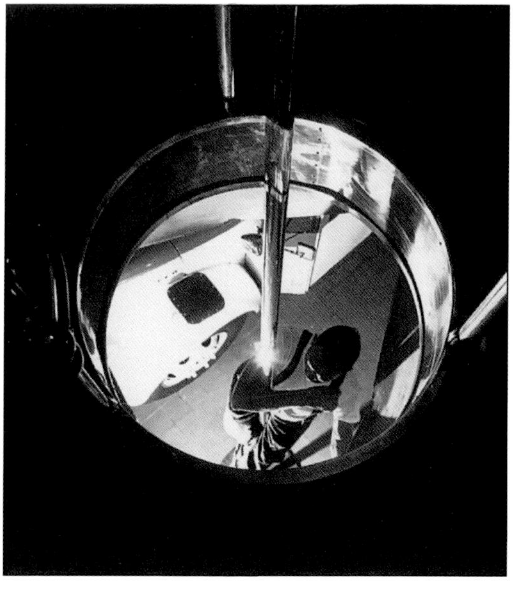

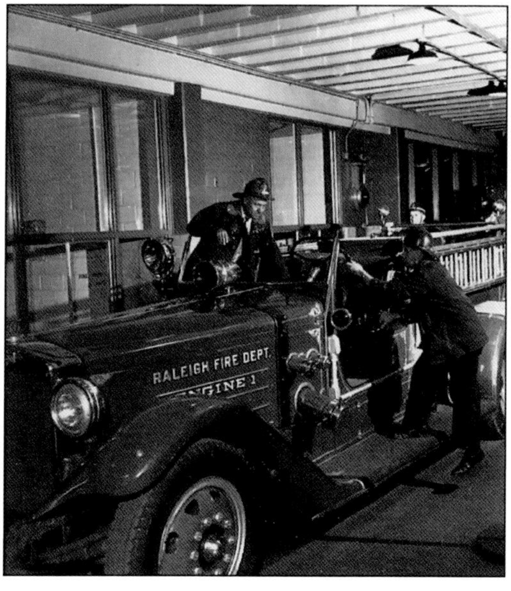

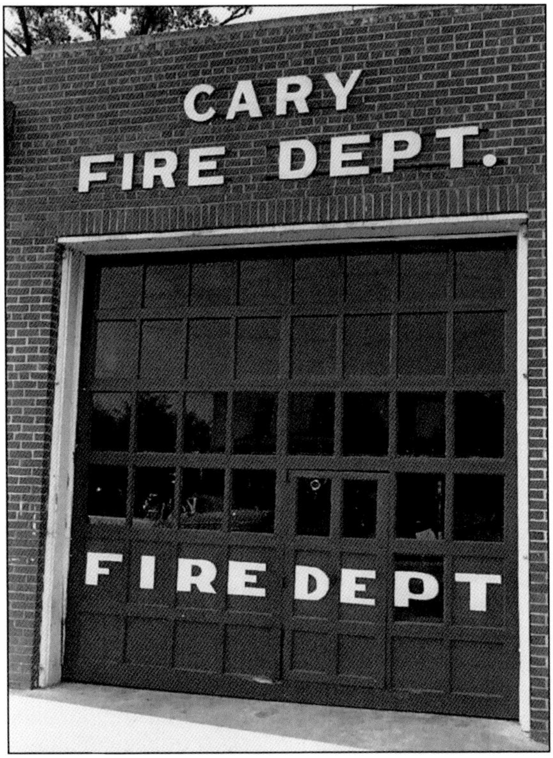

CARY FIRE STATION, JULY 1960. This single-bay building at 100 W. Chatham Street, adjacent to the town hall, housed the fire department from 1953 until 1965. Though only large enough for two trucks, firefighters built a second station on nearby Railroad Street to house additional apparatus. Station 2 was completed in 1954. An earlier fire station was located in a metal garage purchased in 1923 for the town's Ford/American LaFrance chemical car. In 1935, the fire station was moved behind the Masonic Lodge building, now Ashworth's Drugstore. (*N&O* photograph.)

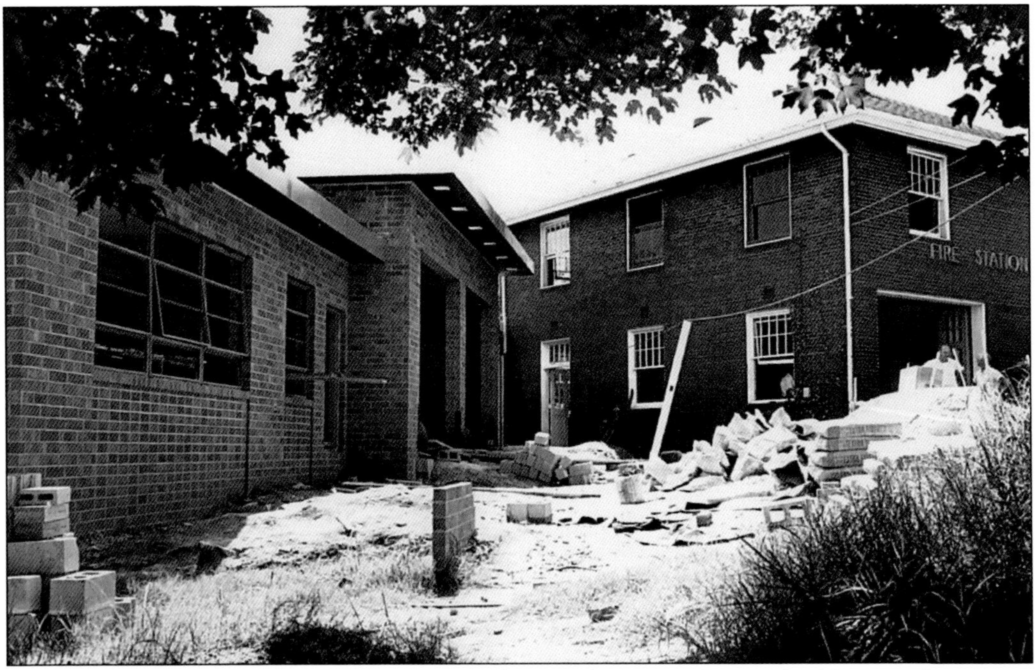

TWO STATIONS, ONE LOT, JULY 1961. Replacing the original 1926 building at the corner of Oberlin Road and Park Drive is a nearly completed Raleigh Fire Station 5. During the construction of the $60,000 facility, Engine 5 was stationed with Engine 6 on Fairview Road. (*RT* photograph.)

SMOKEHOUSE, AUGUST 1964. Wearing breathing apparatus, volunteer firefighters practice at the Raleigh Fire Department smokehouse. Located next to the drill tower, the training building was constructed with bricks removed from old Station 5. Firefighters demolished the old station themselves. The smokehouse remained in operation until the 1980s and was subsequently used as storage. (*RT* photograph.)

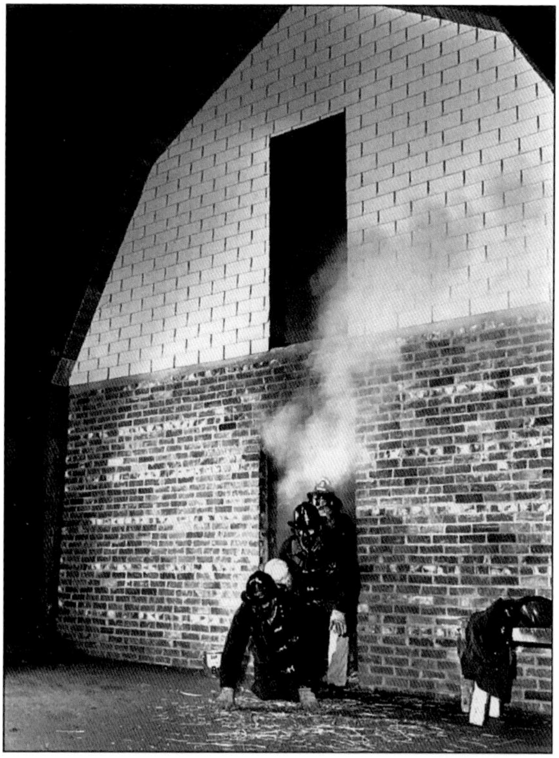

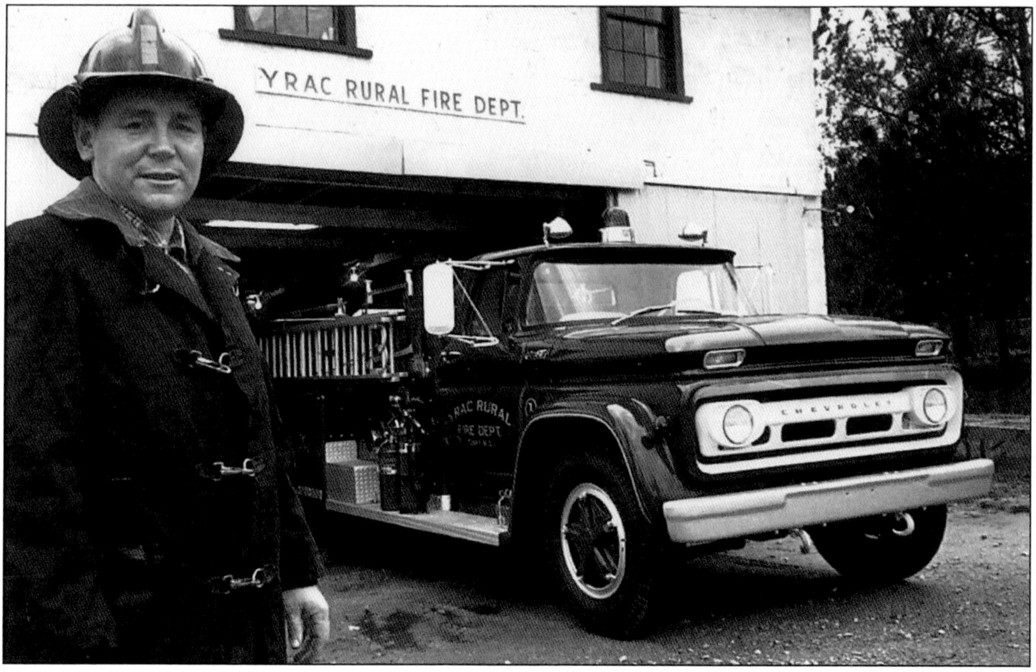

YRAC FIRE STATION, DECEMBER 1964. Cary's rural fire department was first housed in this rented building on Cedar Street behind Cricket's Service Station. William E. Edwards is pictured with their 1962 Chevrolet/American LaFrance pumper. A permanent station was constructed around the corner on E. Durham Road in 1966. (*RT* photograph.)

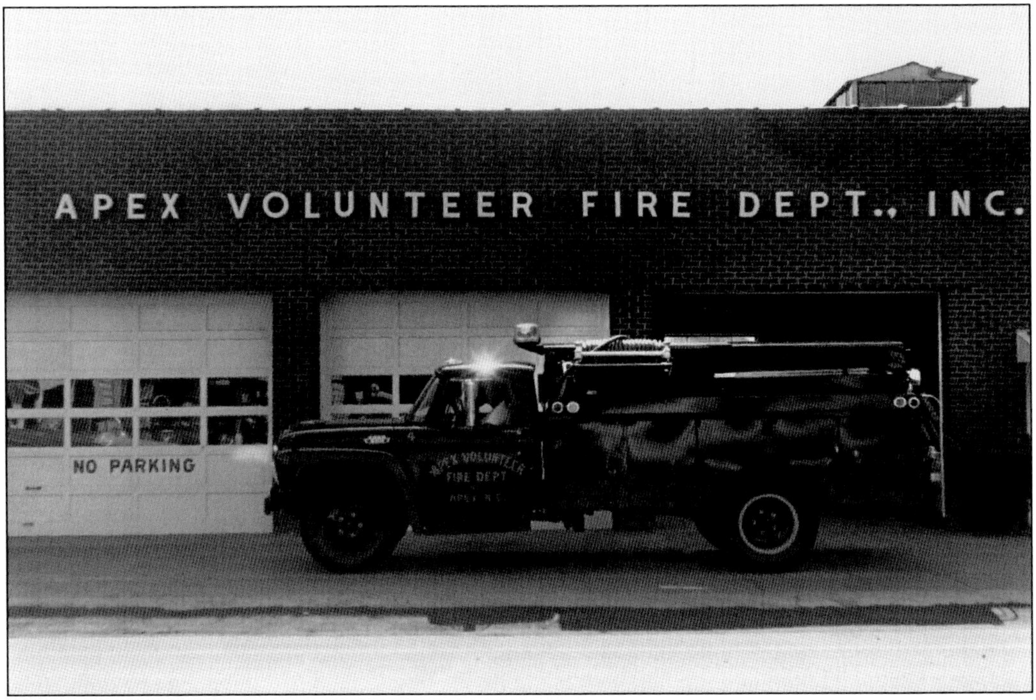

Apex Fire Station, c. 1965. By the time of this photograph, a Civil Defense observation tower had been added to the top of 210 N. Salem Street. Constructed in 1957, the tower was removed when work on a second story was started in 1968. Also pictured is the department's new 1965 Ford foam truck. (AFD.)

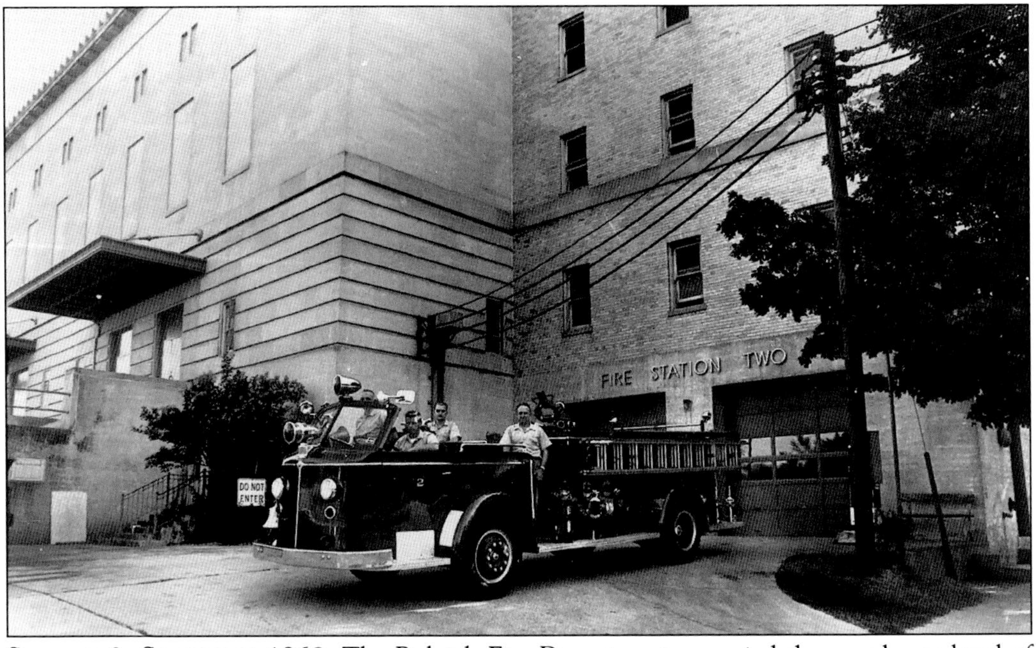

Station 2, September 1968. The Raleigh Fire Department occupied the rear lower level of Memorial Auditorium from 1932, the year it was completed, until 1969, when the fire station was moved to 263 Pecan Street. (*RT* photograph.)

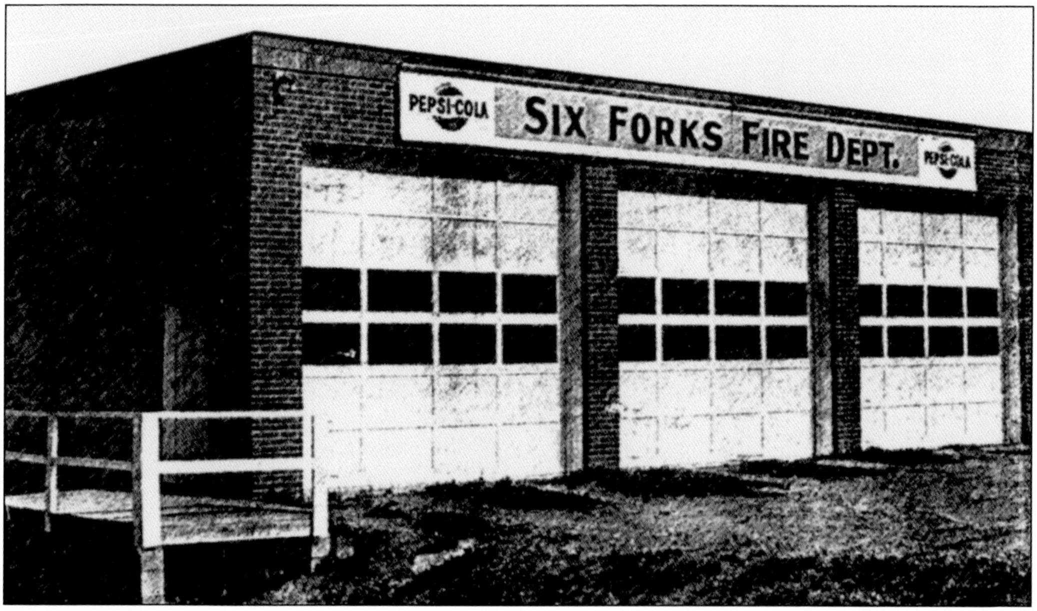

Six Forks Fire Station, 1971. This three-bay building at 5305 Six Forks Road housed the volunteer fire department from 1962 to 1974. Their 7.5 HP siren, however, was located farther north on Six Forks Road at Northclift. The only alarm at the building itself was a 12-volt vehicle siren. In 1974, the fire department (and later its siren) relocated to 1431 Lynn Road. The old building became an EMS station. (Mike Legeros graphic.)

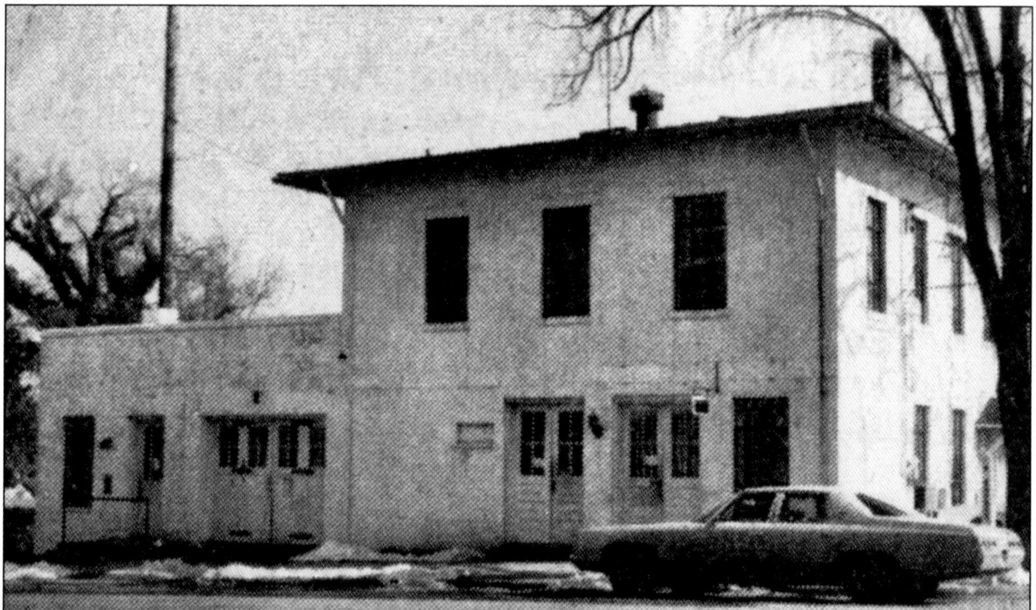

Wake Forest Town Hall, January 1973. Completed in 1930, this municipal building at 221 S. Brooks Street housed Wake Forest Fire Department 1 until it moved into the American Circle Service Station at 350 S. White Street in 1973. Firefighters performed most of the work to convert the service station themselves. (*WW* photograph.)

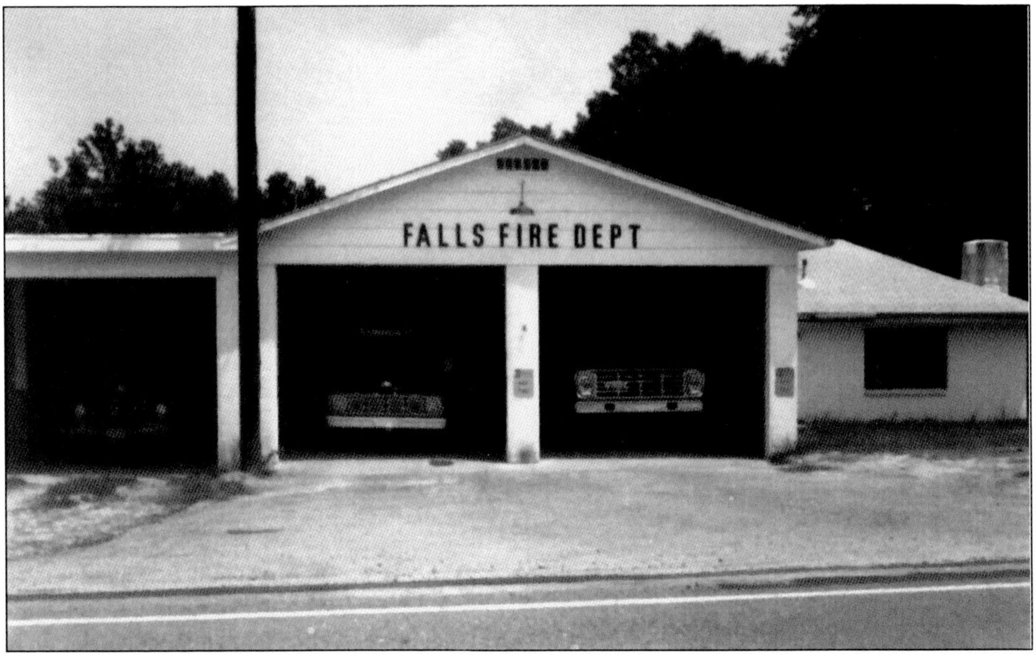

Falls Fire Station. Organized in 1970, the Falls Fire Department built their fire station by adding a pair of apparatus bays onto the Falls Community Center. Two additional bays were added in the 1980s. Later improvements included a paved parking lot, another bay behind the building, and a front bay converted into office space. (FFD.)

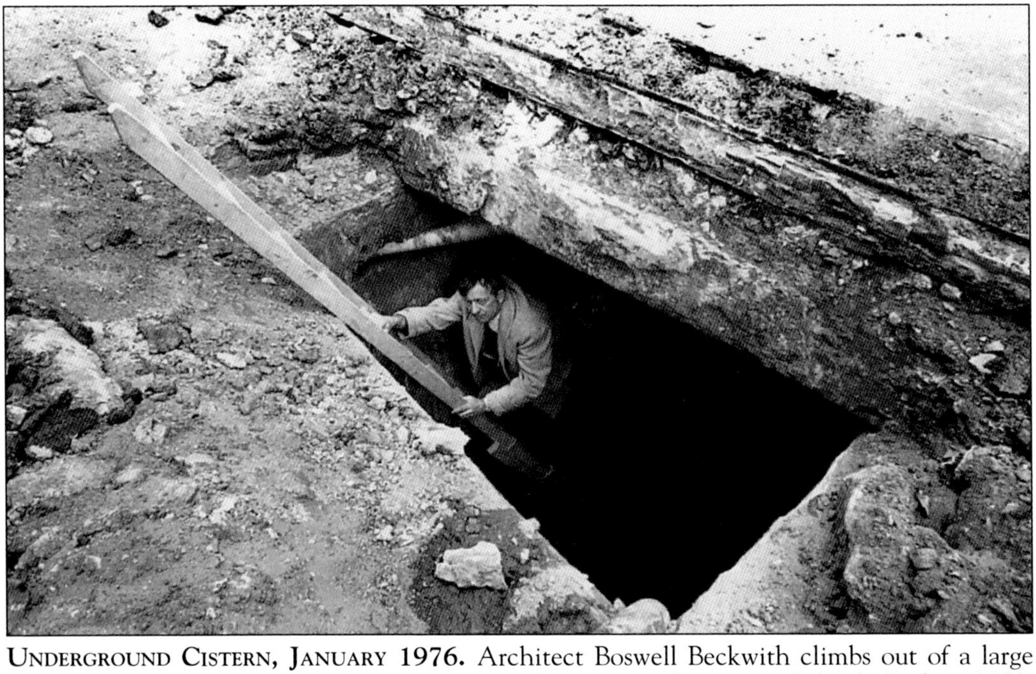

Underground Cistern, January 1976. Architect Boswell Beckwith climbs out of a large brick cistern unearthed by workmen on Fayetteville Street in downtown Raleigh. Built in 1852, the 40,000-gallon tank was one of several used for firefighting before the first hydrants were installed in 1887. (*RT* photograph.)

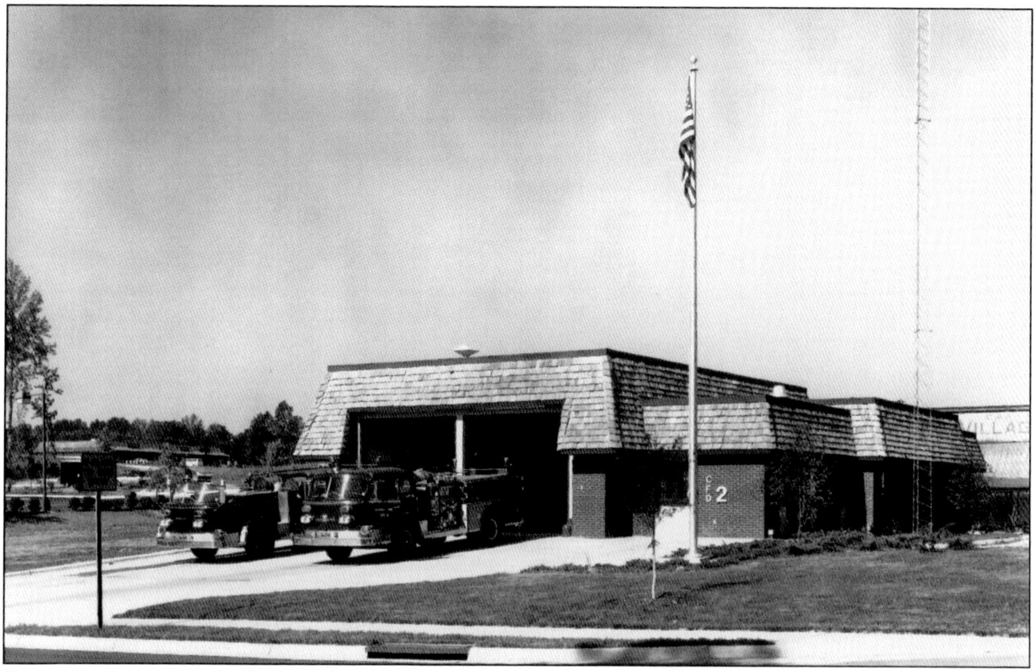

STATION 2, C. 1976. A new Cary Fire Station 2 was completed that year at 875 NE Maynard Road. The $185,000 building is shown with Engine 1 and Engine 2 posed on the apron. Four years later, it also housed a sign shop for creating and maintaining the town's street signs complete with a lamination machine built by firefighters. (CFD.)

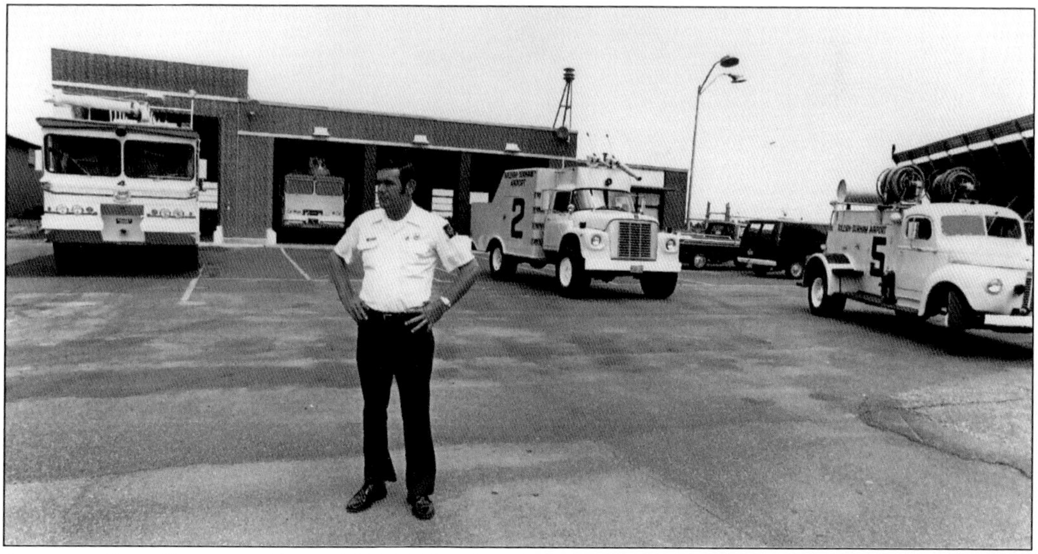

AIRPORT FIRE STATION, JUNE 1977. This four-bay fire station was located at the north end of the South Ramp at Raleigh-Durham Airport. Originally built with three bays and an office, a fourth larger bay was added in 1973. The fourth bay also had a classroom in the rear. To the right is the blast shield that protected the general aviation area from prop and jet wash. Fire Chief Terry Edmondson is pictured with, from left to right, a 1973 Walter CB3000, a 1959 Walter Class 1500, a 1969 International/Ansul Mangum 480, and a 1946 International/Bean. (RT photograph by Karen Tam.)

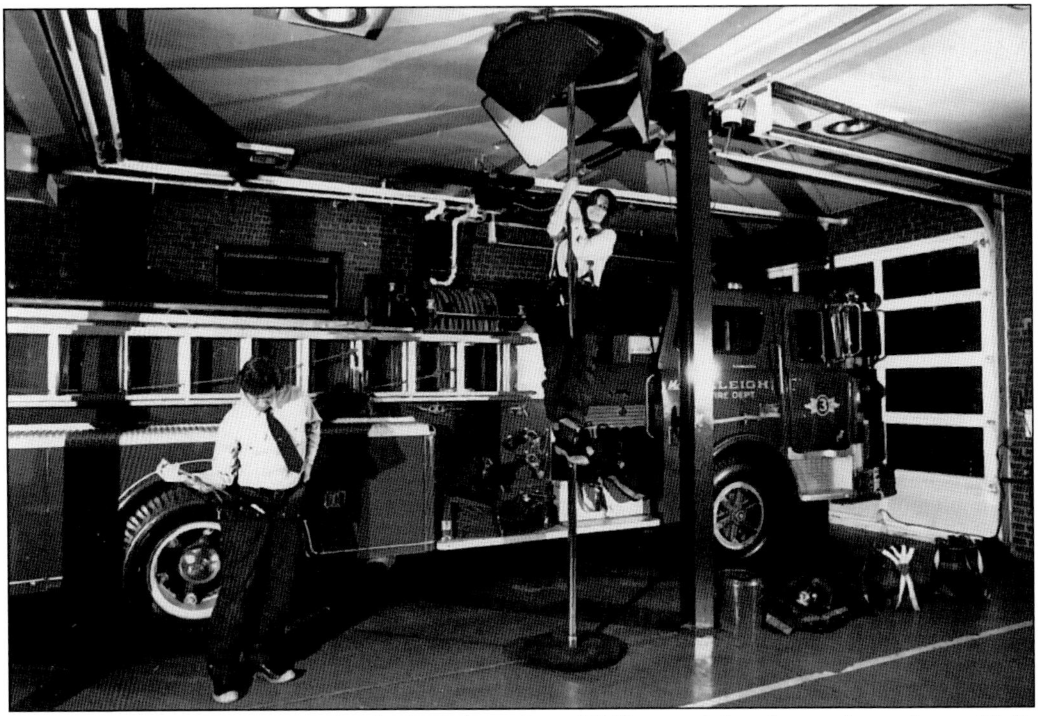

Station 3, April 1980. Raleigh firefighter Pam Williams slides down the pole at 13 S. East Street for a *News & Observer* feature on female firefighters. Williams and seven other women were hired in July 1978, the first in the fire department's history. (*N&O* photograph by Gene Furr.)

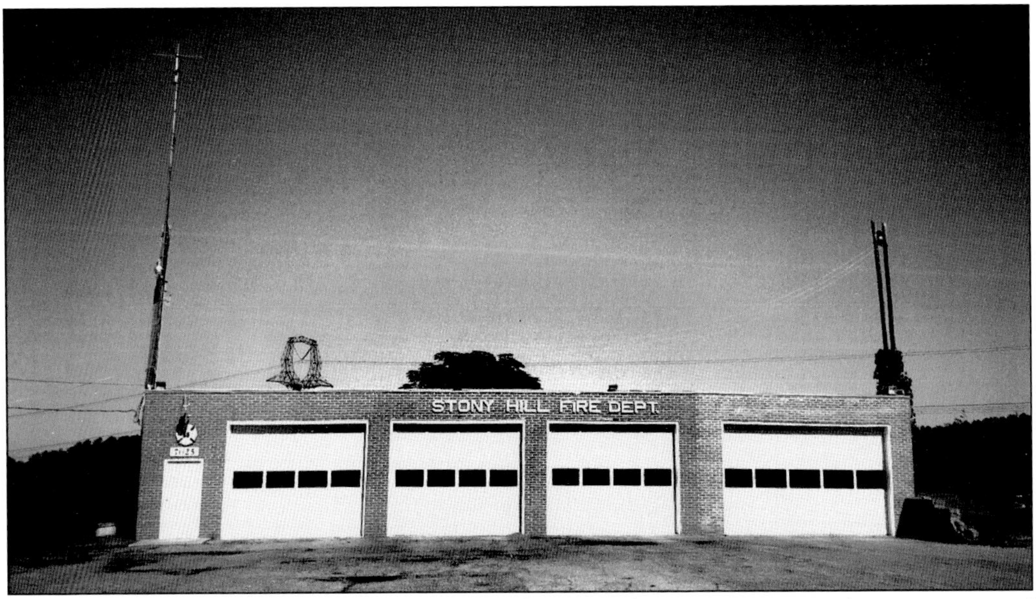

Stony Hill Fire Station. In 1963, firefighters and community members constructed this permanent fire station at 7025 Stony Hill Road. Built on land loaned by Bill Ray, the three-bay building was expanded in 1981 to include a fourth bay and a meeting room. (Mike Legeros photograph.)

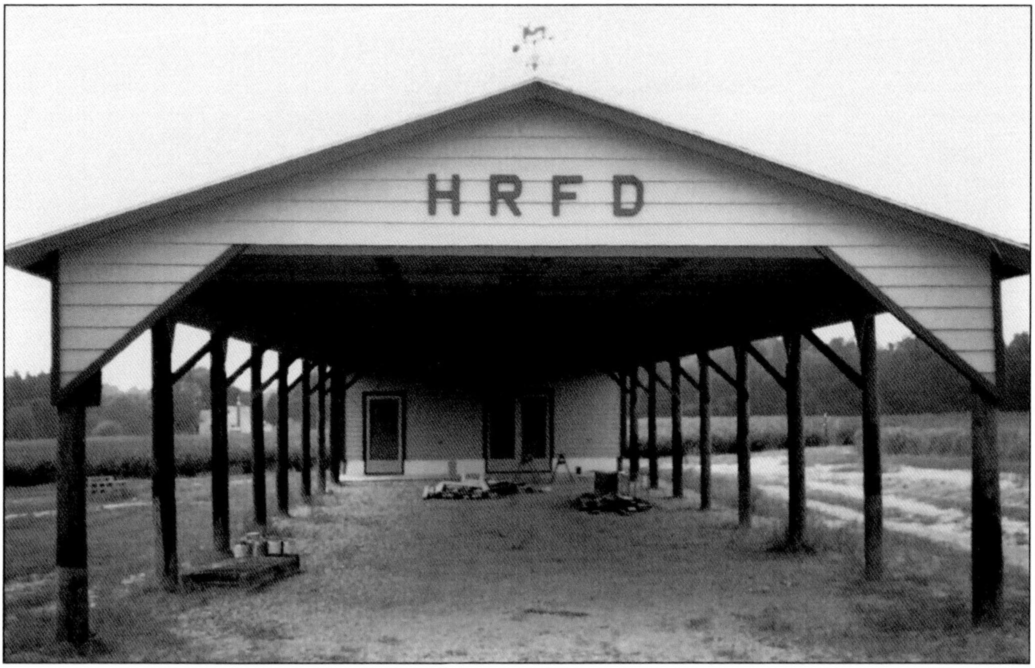

SHELTER, AUGUST 1982. Groundbreaking for the Hopkins Fire Department's picnic shelter was held on February 23, 1982. The 120-foot long structure was built by firefighters and community members with both monetary and material donations, including trees for lumber and "50 pounds of nails" from Mr. Frank Smith. (HFD.)

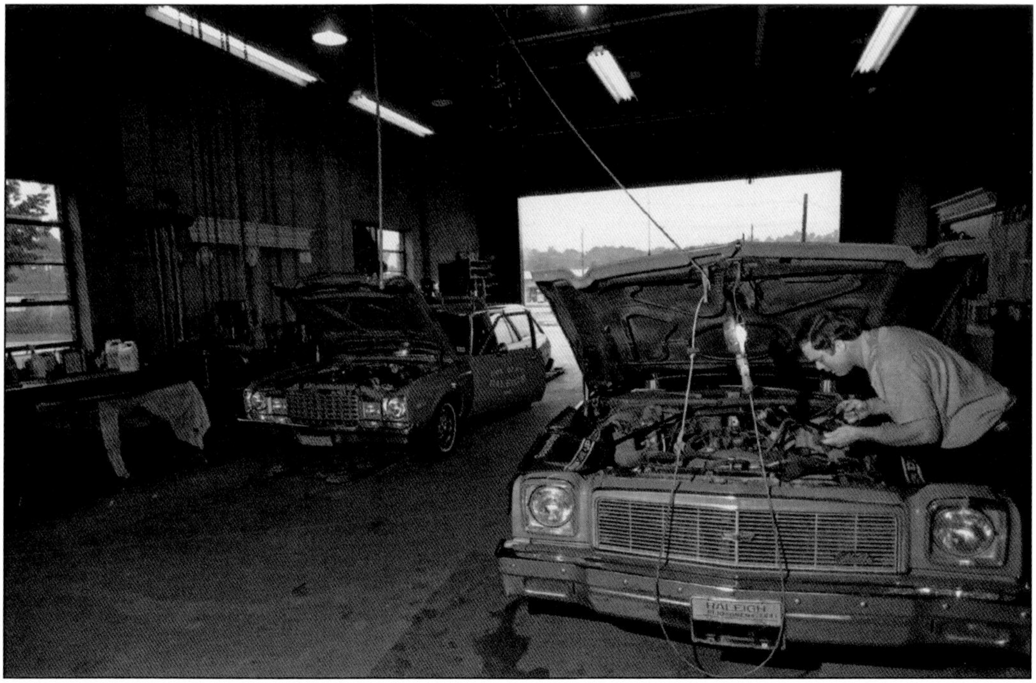

SHOP, 1984. Mechanic Richard Spangler is pictured at the Raleigh Fire Department's maintenance facility located behind Station 2 at 263 Pecan Street. (Gary Knight/CCBI photograph.)

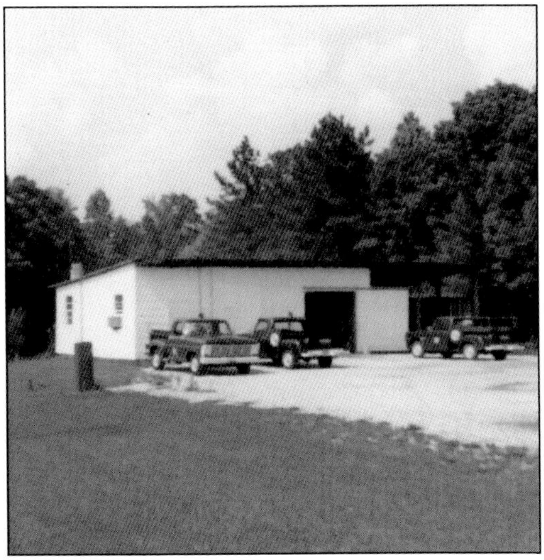

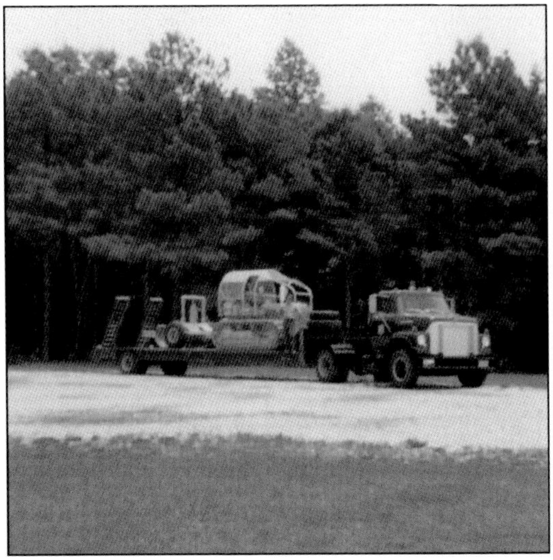

FOREST SERVICE, 1982. This building below the Bay Leaf fire tower at 2600 Howard Road housed ranger offices, a weather station, a maintenance area, and a shelter for the tractor-pulled plow, pictured at right. (Don Adams photographs.)

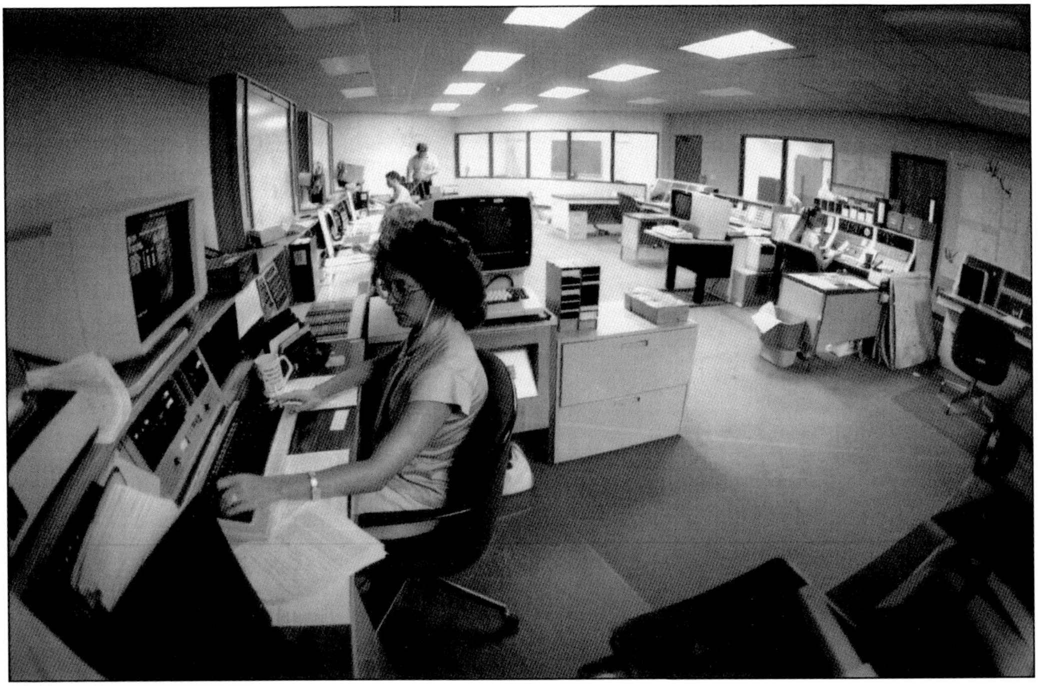

COMMUNICATIONS CENTER, C. 1984. Beginning in 1972, the new Raleigh/Wake County Emergency Communications Center assumed dispatching duties for nearly all of Wake County's fire departments. Shift supervisor Jane Diedrich is pictured in their second location, beneath the municipal building at 222 W. Hargett Street. The center was originally located next door in the old municipal building at 110 S. McDowell Street. (Gary Knight/CCBI photograph.)

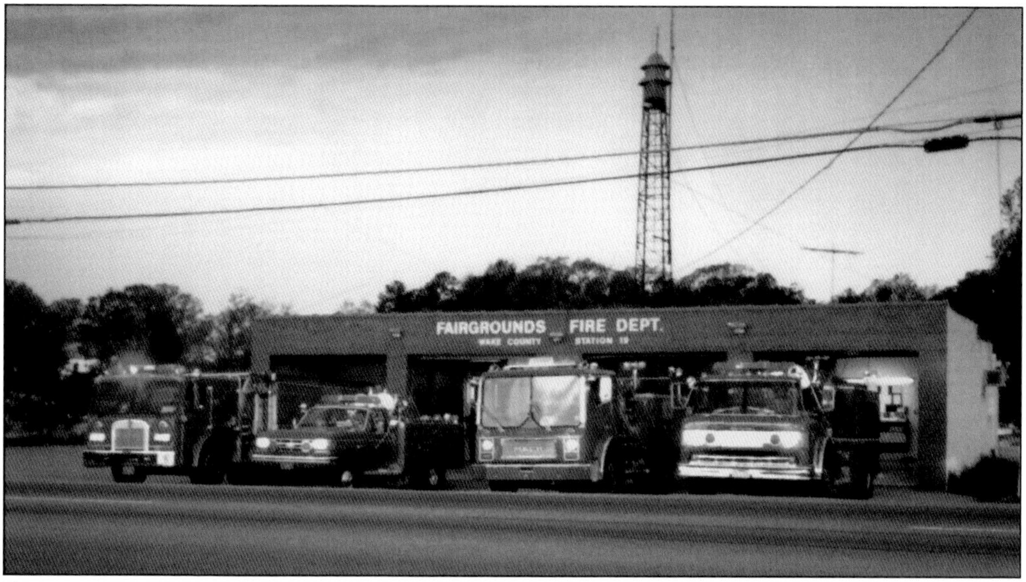

Fairgrounds Fire Station, April 1989. These cramped quarters at 1515 Blue Ridge Road housed the Fairgrounds Fire Department until 1992. Originally a two-bay building constructed in 1961, the interior was renovated in 1968–69 after an overflowing oil heater started a fire. Two additional bays and a day room were added in the mid-1970s. The latter later served as a sleeping area for the increasing numbers of college students volunteering in the late 1980s. North Carolina State University students slept on chairs, couches, and even the apparatus itself. The jump seat of Engine 196 and the hose bed of Tanker 198 were particularly comfortable. Pictured above, from left to right, are a 1985 Kenworth/Grumman Pumper-Tanker, a 1987 Ford/Grumman brush truck, a 1981 Mack Pumper, and a 1985 Ford/Grumman Tanker. Picture below, left, is a 1988 Chevrolet suburban. (WWFD.)

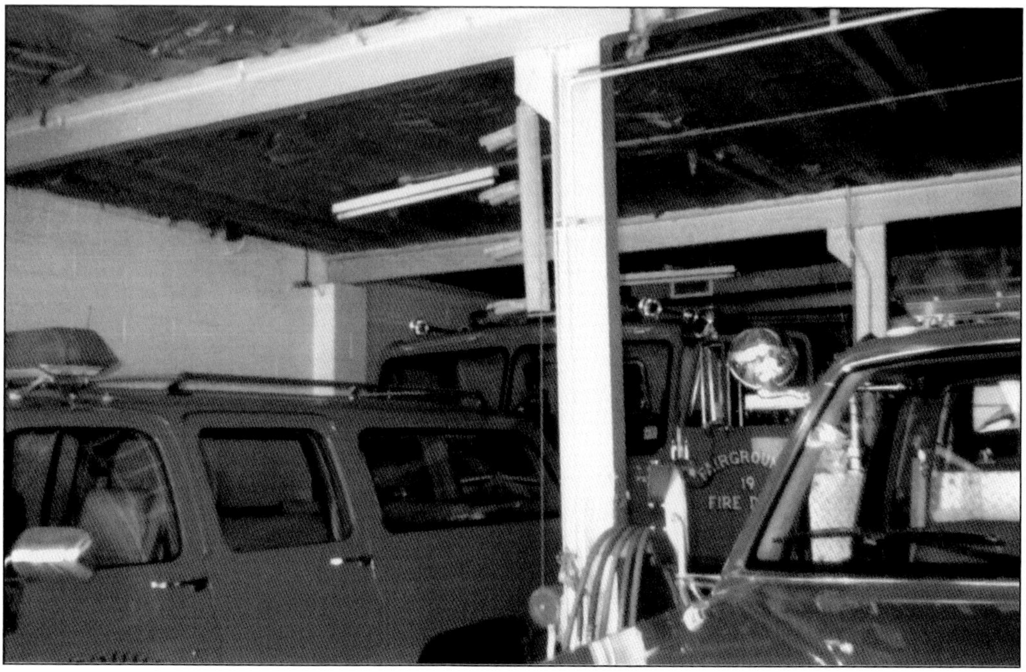

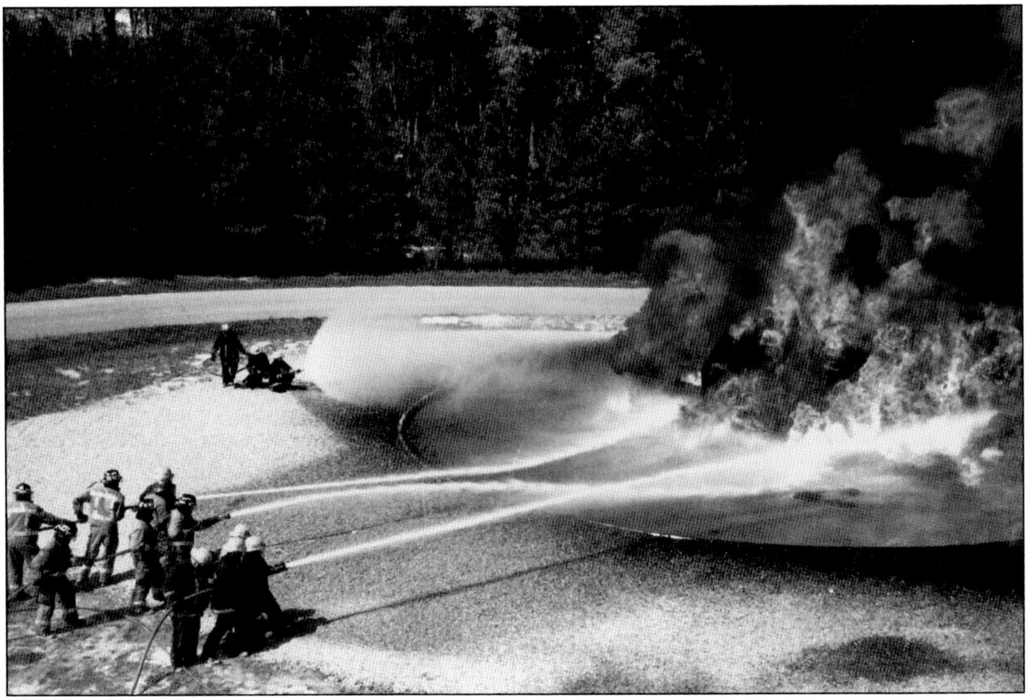

TRAINING CENTER. Located on eight acres in New Hill, the Wake County Fire Training Center was built in 1985 and first included a burn pit and a burn building. Later additions to the $700,000 facility included a five-story drill tower, propane exercises, and a flashover simulator. Full use of the facility began in September 1988 with the start of a countywide training program. By the end of 1989, after the Wake County Fire Academy's first full year of operation, 11,010 man-hours of live burns had been conducted. (Courtesy of Wake County Public Safety Fire/Rescue Division.)

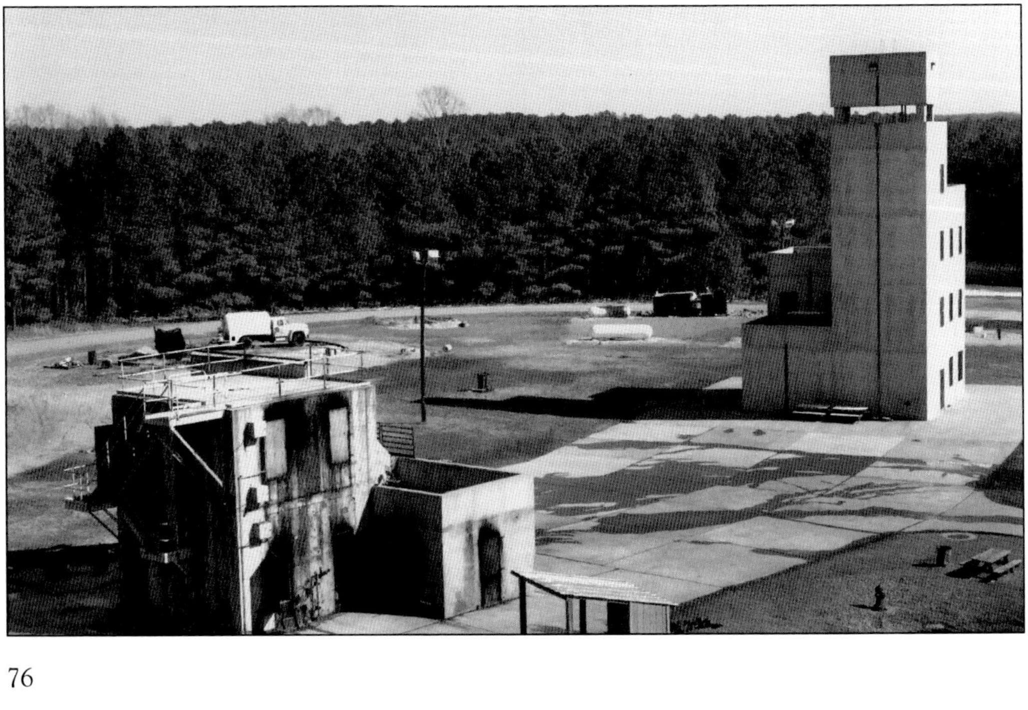

Four

Personnel

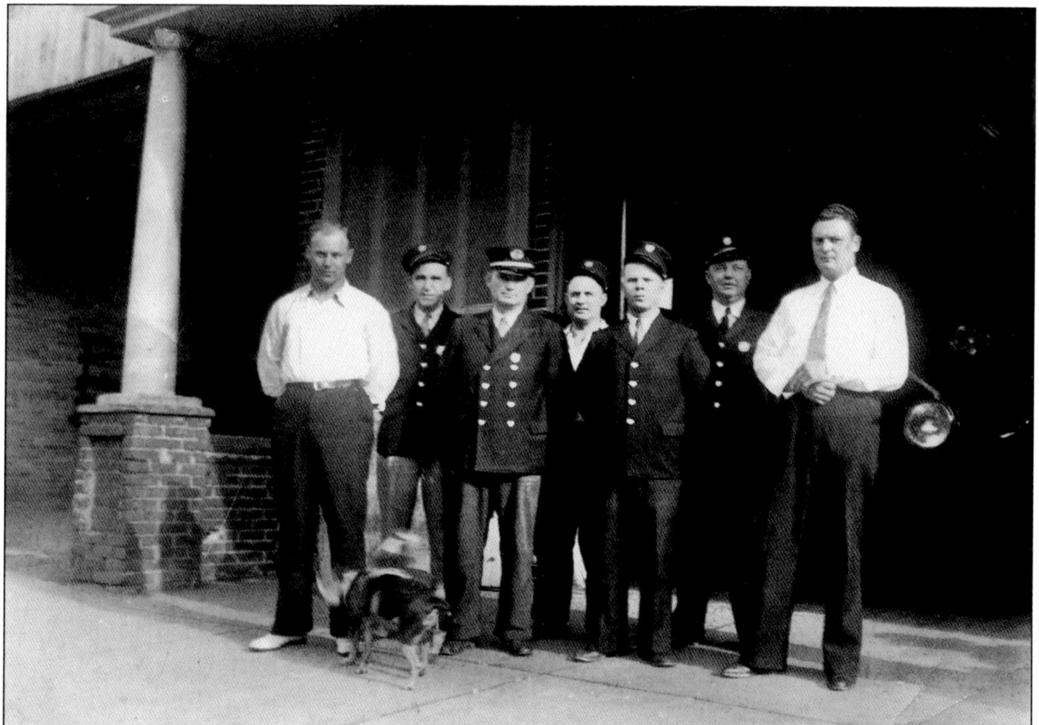

Station 3, 1943. Standing in front of 135 E. Hargett Street, from left to right, are Raleigh firemen Bill Lindsay, Lawson "Foots" Glenn, Capt. Leonard Choplin, Jimmy Poole, B.C. Dowd, Dempsey Collins, and Bill Kelly. (Courtesy of Betty Poole Brinkley.)

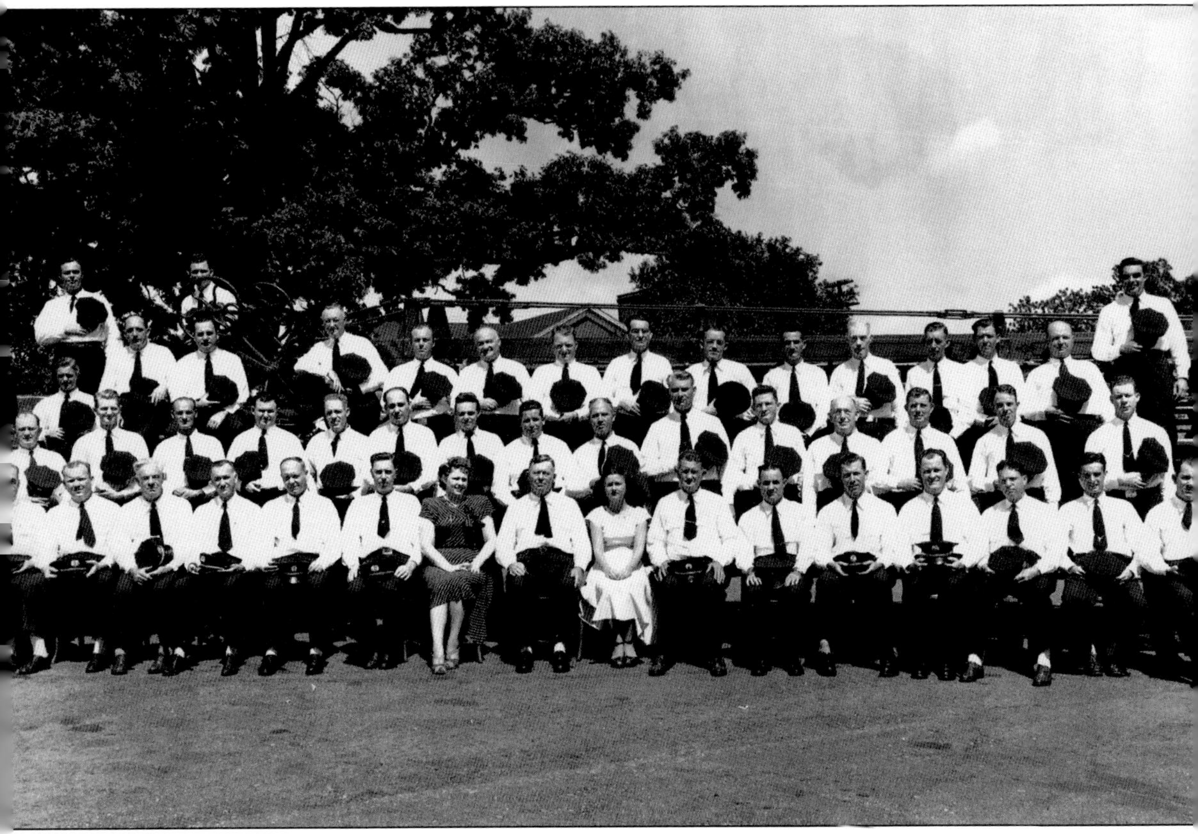

Raleigh Fire Department, c. 1948. Pictured, from left to right, are (front row) Capts. James M. Burnette, Royce C. Lassiter, and John G. Harrison; Lt. Horace N. Sadler; Capts. Edward G. McGhee and Jack B. Keeter; Delia Isaacs; Fire Chief Alvin B. Lloyd; Irene Young; Asst. Chief R. Lee Matthews; Lts. James A. Poole, David L. Brannan, James T. White, and Ulysses M. Ennis; Pvt. Glenn S. Harrison; and Driver Julius C. Hodge; (second row) Driver May T. Parker; Pvt. John W. Gregory; Driver Floyd T. Pipkin; Pvt. H. Bagwell Williams; Drivers Charles E. Nunn and Clarence R. Puryear; Pvts. Roy T. Hamlet and James E. Carroll; Driver Virgil G. Mims; Pvts. Lucius Q. Godwin, Harold S. Stephenson, and Moses E. Perry; Driver Roy R. High; and Pvts. William L. Mitchell and Ernest C. Emory; (third row) Drivers Carl L. Wall, John W. Godwin, Vernon J. Smith, Jack C. Crabtree, J. Theodore Honeycutt, and J. Robert Marshall; Pvt. Dennis B. Grubbs; Driver Joesph S. Allen; Pvt. Brantley B. Olive; Drivers Oscar Summers, Boaz A. Edwards, and Joesph M. Hobby; Pvt. Eugene J. Alford; and Driver Dempsey D. Collins; (standing on truck) Pvts. John W. Holmes, James E. Fowler, and A. Eric O'Neal. Not present are Capt. Charles Hayes and Pvt. A. Wade Watkins. In 1948, the Raleigh Fire Department had six engine companies, two ladder companies, and an auxiliary truck. They responded from six stations located at 412 S. Salisbury Street (Station 1), 735 Fayetteville Street at Memorial Auditorium (Station 2), 135 E. Hargett Street (Station 3), 505 Jefferson Street (Station 4), 1914 Park Drive (Station 5), and in a rented building at 2513 Fairview Road (Station 6). (Courtesy of C.T. May.)

PAINTING, MARCH 1949. Raleigh firefighters completely overhauled the interior of Station 5 at 1914 Park Drive, including building a new floor, kitchen, and bathroom. Putting on the finishing touches are, from left to right, Oscar Summers, Henry Proctor, Capt. Ulysses Ennis, and Bill Mitchell. Firemen performed almost all repair and remodeling work in the "old days." City Council minutes dated January 2, 1951, recount the approval of an additional room at Station 4 for use as a kitchen. Firefighters agreed to perform the work if the city provided $1,174 in materials. (*RT* photograph.)

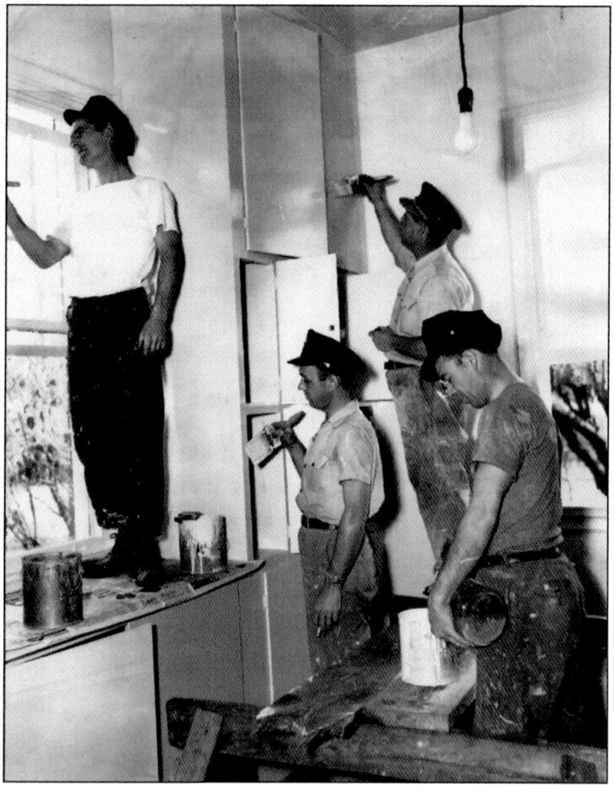

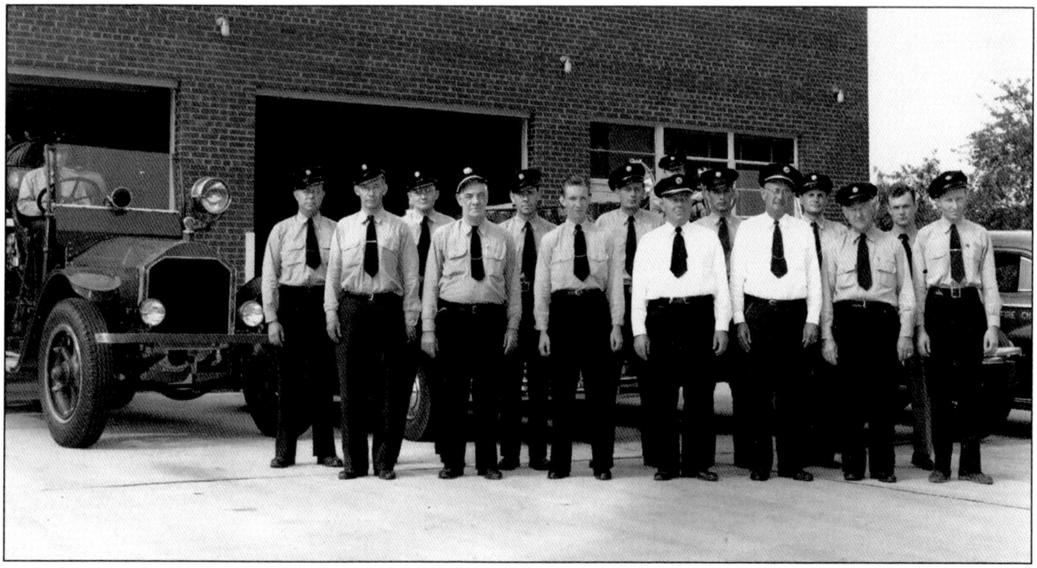

STATION 6, 1950. Pictured from left to right at 2601 Fairview Road are (front row) Joe Hobby, Capt. John Harrison, Clyde Carter, Fire Chief Alvin Lloyd, Asst. Chief Lee Matthews, Capt. Jack Crabtree, and Bill Durden; (back row) Roy High, Harold Stephenson, Eugene Alford, James Strickland, Thomas Gates, John Todd, and Frank Beacham; (sitting in apparatus) Vernon Smith and Wade Watkins. (Courtesy of Clyde Carter.)

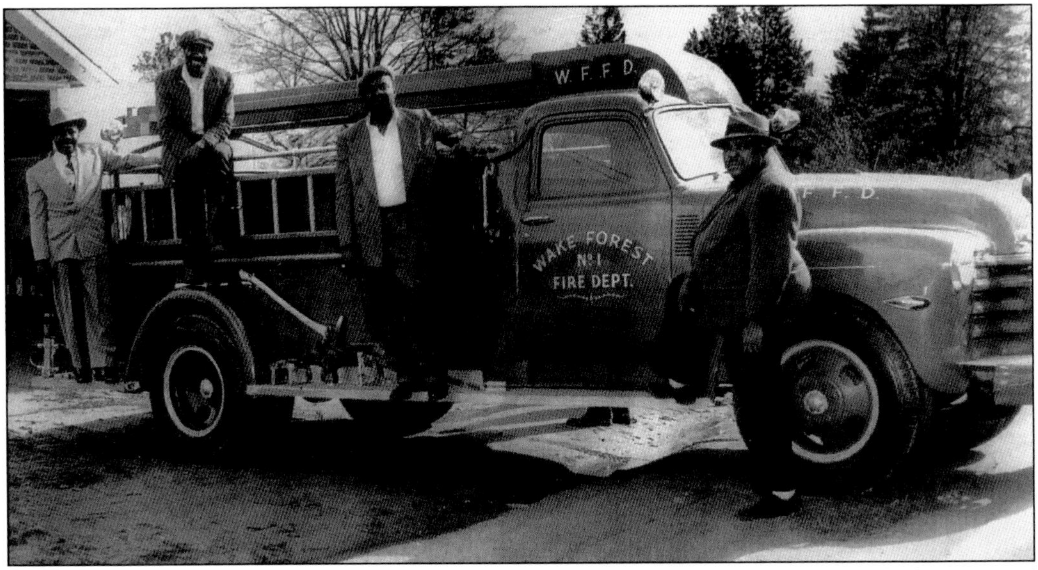

WAKE FOREST FIRE DEPARTMENT 2, c. 1953. Pictured with their 1947 Chevrolet/American/Barton pumper are, from left to right, Fire Chief Edward Alston, Matthew Williams, George Massenburg, and Robert Alston. (Courtesy of Wake Forest College Birthplace Society.)

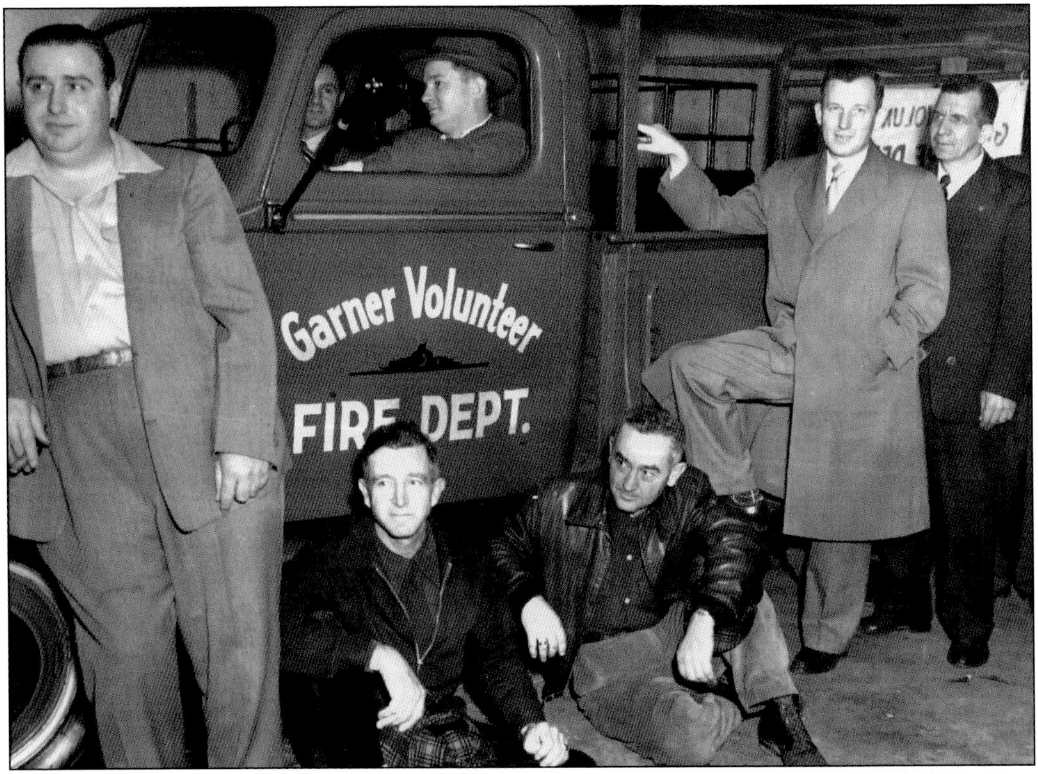

GARNER FIRE DEPARTMENT, c. 1953. Pictured with their first fire truck are, from left to right, (front row) Joe Buffaloe, Sam Carroll, J. Noel Bryan, Charlie Ridoutt, and Raymond Umstead; (in truck) Horace Barrow and Archie Castleberry. Before they built the firehouse in 1953, the fire engine was stored in a garage on Garner Road across from the high school. (GFD.)

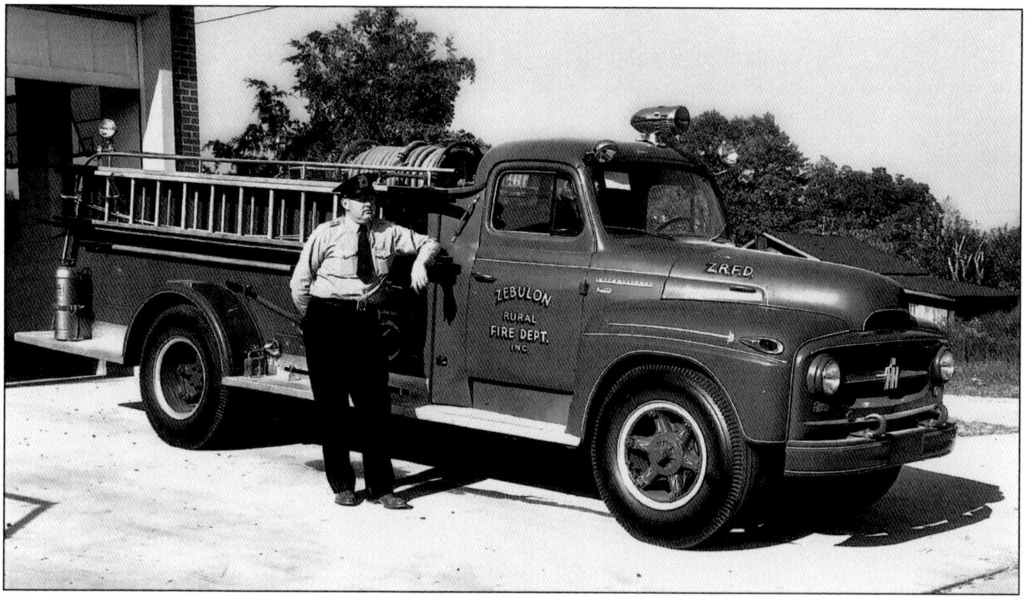

WILLIE B. HOPKINS, OCTOBER 1953. The Zebulon police chief poses with the new truck of the new rural fire department. Hopkins, also the chief of the town fire department, later served as Clerk of Court, Tax Collector, Water Works Superintendent, and Town Manager. (*N&O* photograph.)

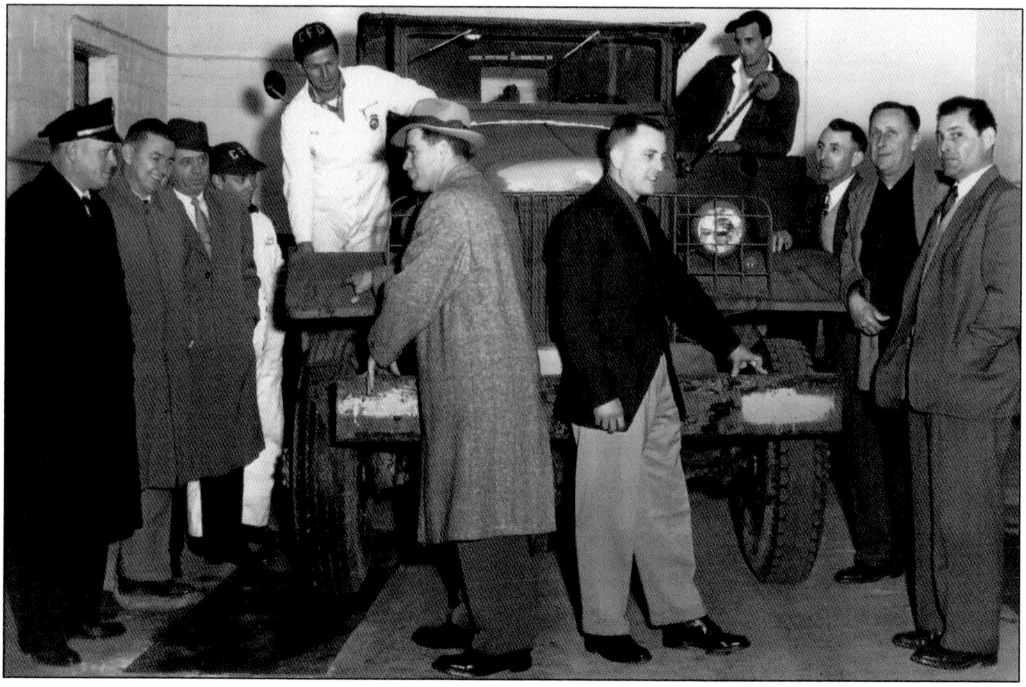

CARY FIREFIGHTERS AND TOWN COMMISSIONERS, MARCH 1954. Inspecting an old truck purchased for conversion into a fire engine are, from left to right, commissioners L.E. Midgette, Billy Creel, and Raymond B. Morgan; firefighters Billy Henderson, J.C. Griffis, and Bob Elder; Fire Chief James L. Murdock; firefighter Preston Wrenn; commissioner Jimmy Hogarth; Mayor H. Waldo Rood, and commissioner Frank Blackley. (*RT* photograph.)

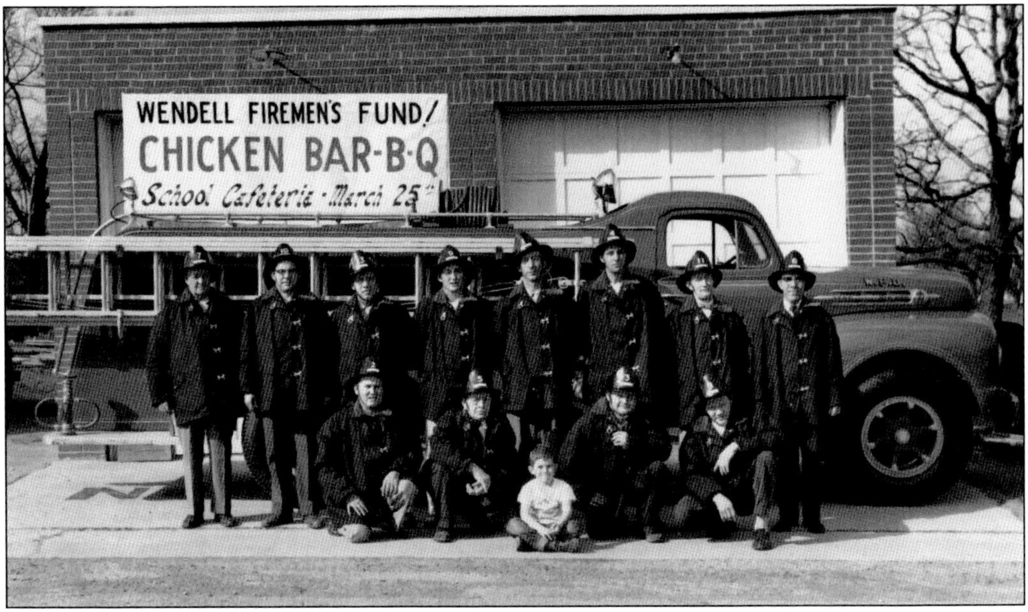

Wendell Fire Department, c. 1954. Pictured in this photograph by John Mattox are, from left to right, (kneeling) Ransom "Pete" Johnson, J. Malcomb Todd, Doug Mattox, C. Proctor "Proc" Dean, and Clifton "Red" Couick; (standing) Colin Doan, C. Bailey Scarboro, Hugh Moody, Richard Frady, Dover Hinton, Bill Conoley, Ira Johnson, and W. Crawford Coley. (Courtesy of Buddy Scarboro.)

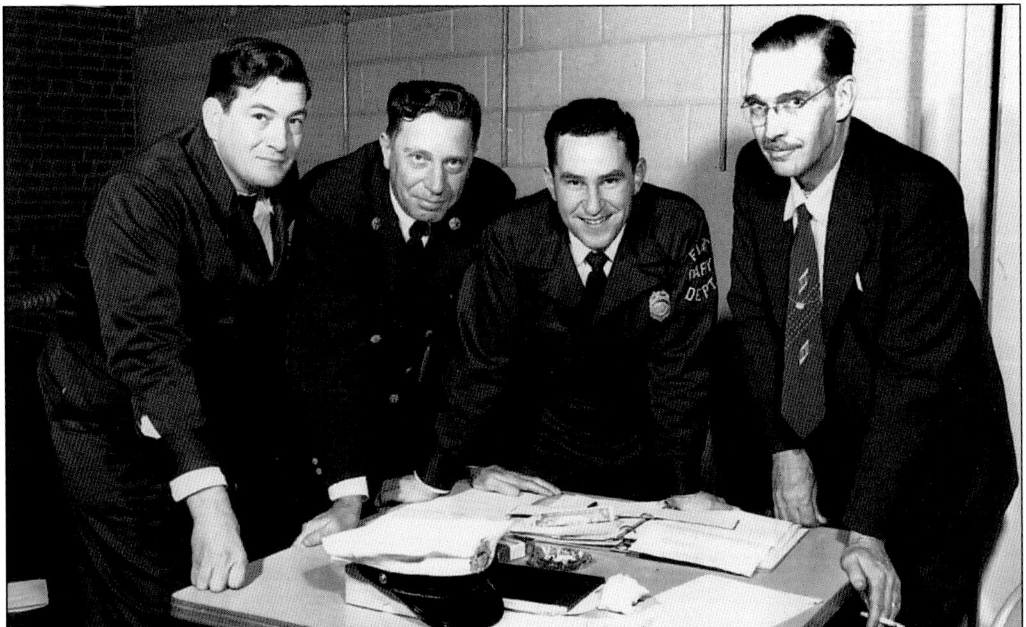

Wake County Fireman's Association, January 5, 1956. This association was chartered on November 23, 1955, and its first officers, from left to right, were secretary-treasurer Gordon Keith Jr. of Cary, first vice president Jack Keeter of Raleigh, president Bob Heater of Cary, and second vice president Carter Schaub of Apex. (*N&O* photograph.)

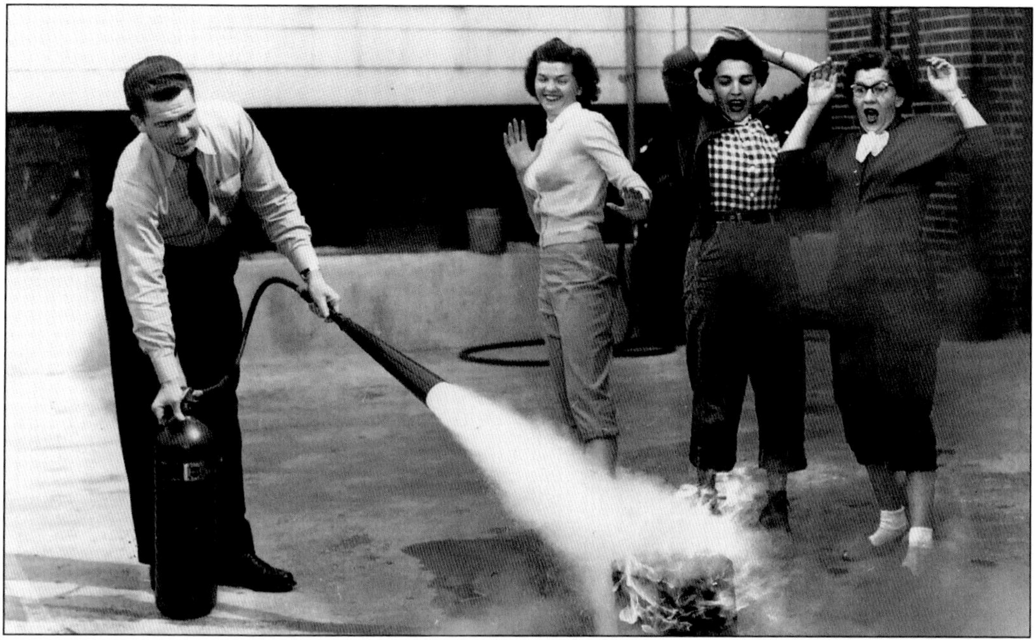

FIRE EXTINGUISHER, APRIL 1956. Raleigh firefighter Harold Jones conducts a fire prevention class for housewives. His excited pupils are, from left to right, Mrs. C.M. Gillispie, Mrs. Paul Pendergraft, and Mrs. George Hunter. (*RT* photograph.)

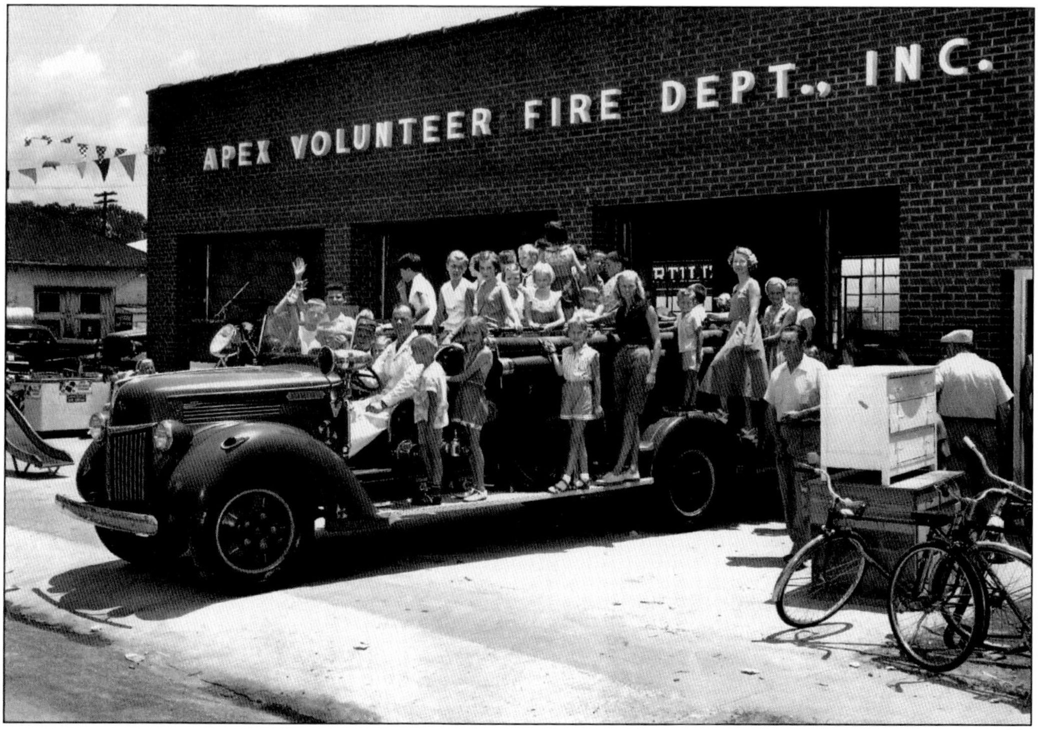

APEX FIREMAN'S DAY, JULY 4, 1956. Children climb aboard the town pumper, one of two trucks assigned small haulage as part of the third annual Fireman's Day festivities. Other activities included a parade and dedication of the new fire station. (*RT* photograph.)

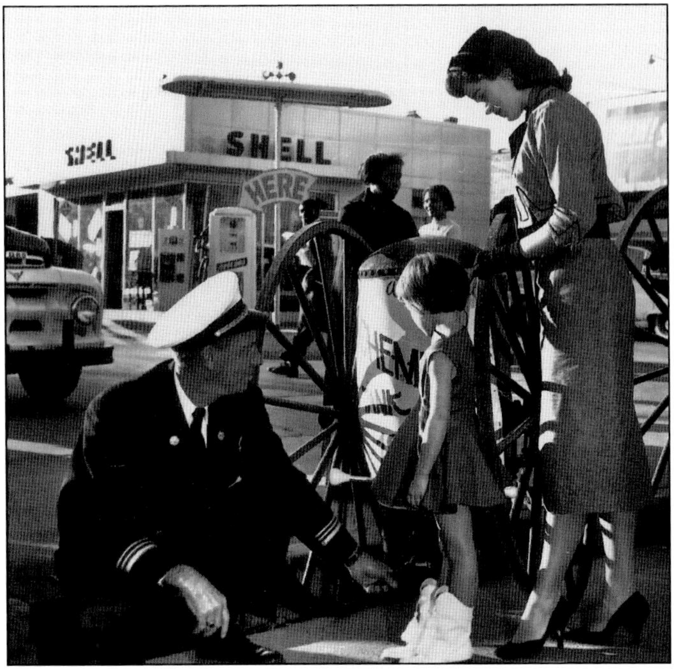

YOUNG AND OLD, OCTOBER 1956. Raleigh fire chief Jack Keeter shows off a 19th-century hand-drawn chemical cart. The Phoenix Chemical Company, one of Raleigh's later volunteer fire companies, operated both a single- and double-tank Champion "chemical engine." (*N&O* photograph.)

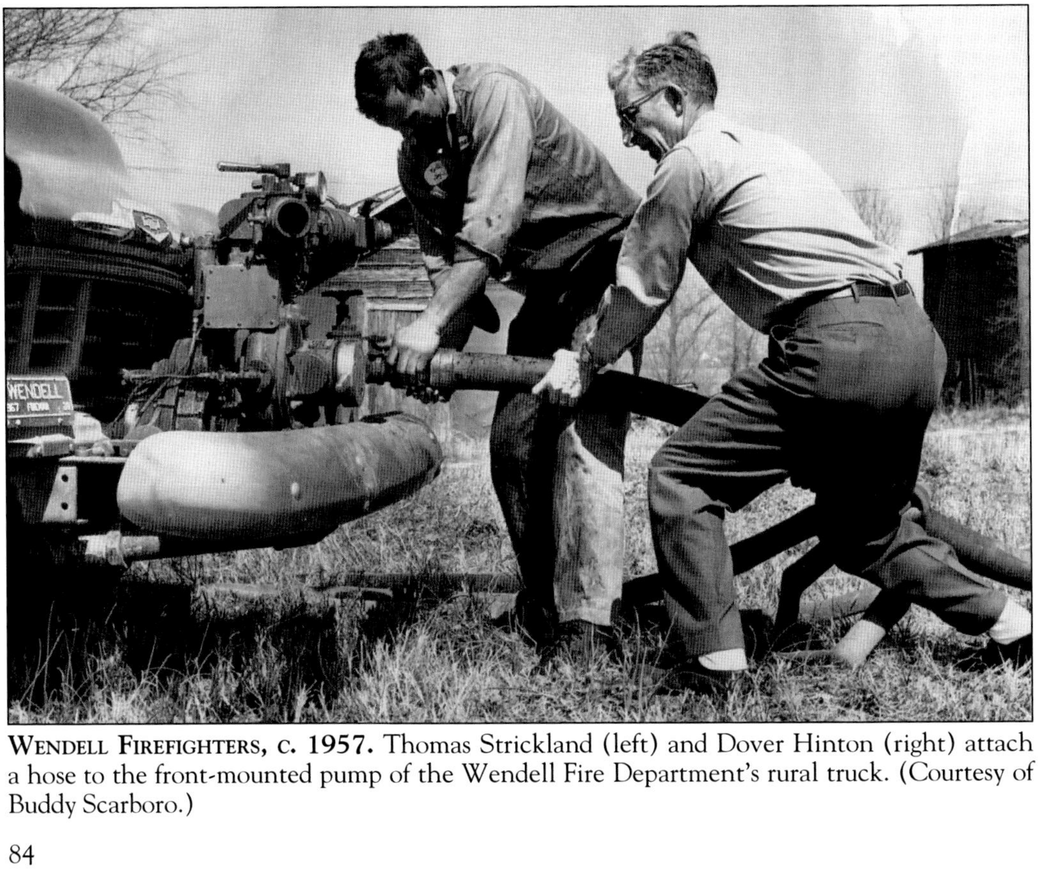

WENDELL FIREFIGHTERS, C. 1957. Thomas Strickland (left) and Dover Hinton (right) attach a hose to the front-mounted pump of the Wendell Fire Department's rural truck. (Courtesy of Buddy Scarboro.)

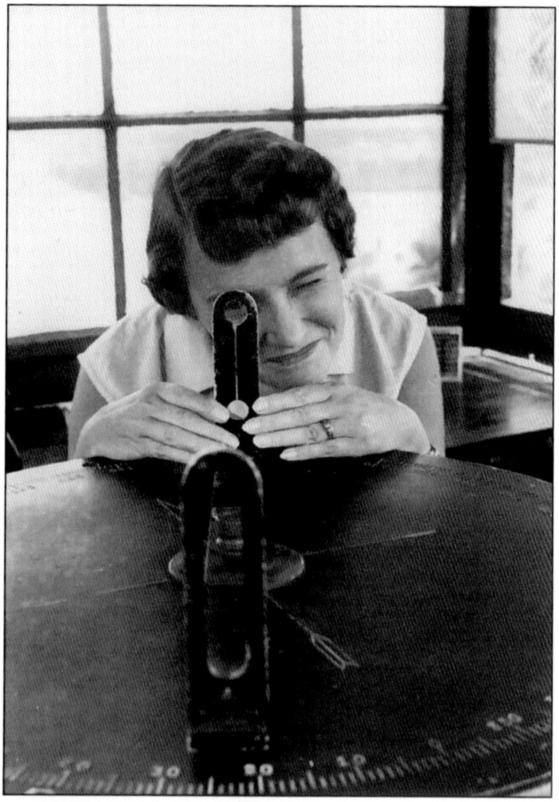

Fire Spotter, May 1957. From the top of the Bay Leaf fire tower, Jenny Tilley demonstrates using an alidade to pinpoint the location of a forest fire. The combination telescope and compass was the tool of her trade, along with a telephone and, in later years, a two-way radio. Built in 1936 on the west side of Six Forks Road about a half-mile north "Six Forks," the 99-foot, 9-inch Aerometer steel tower was moved to its present Howard Road location in 1967. Tilley, wife of Wake County Forest Ranger George Tilley, manned the tower at both locations until its use was discontinued in 1985. On a clear day, she could see some 25 miles. (*N&O* photograph by Burk Uzzle.)

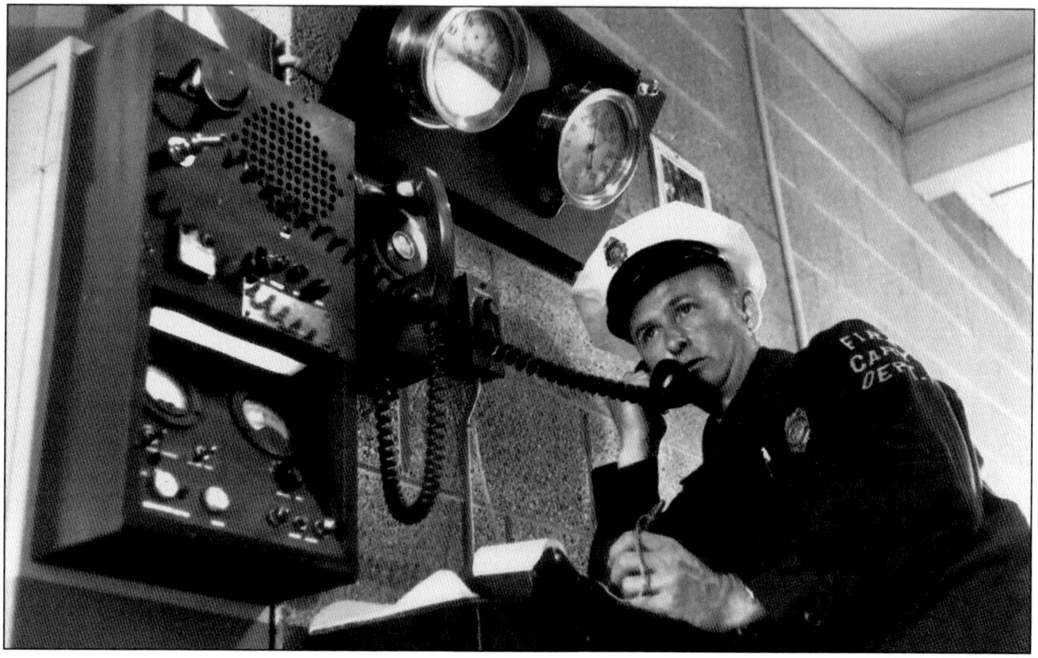

Radio Equipment, April 1958. Cary fire chief J. Paul Matthews demonstrates the department's new two-way radio equipment. The 500-watt network connected the fire station with each of the fire engines, the rescue truck, and the fire chief's car. (*RT* photograph.)

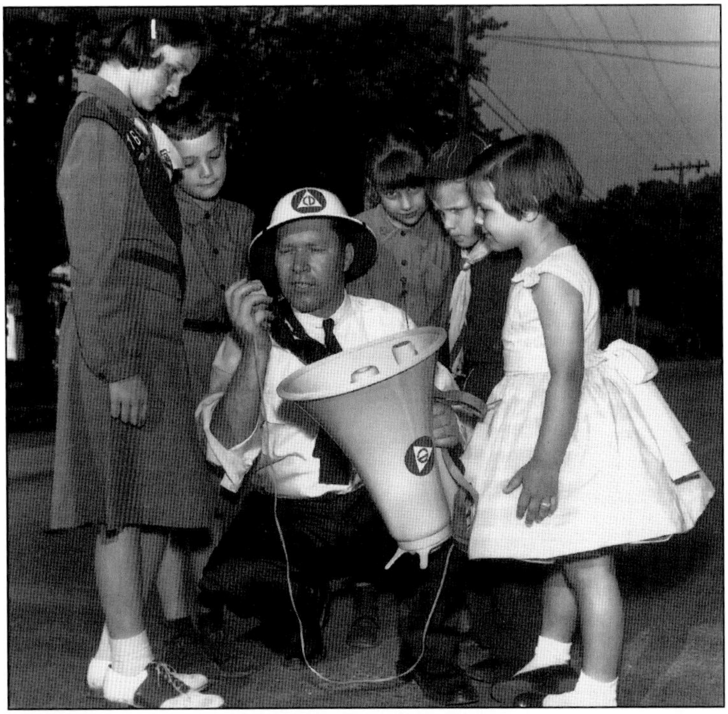

CARY FIREMAN'S DAY, MAY 1958. Watching firefighter George "Buck" Sloan talk into a loudspeaker are, from left to right, Ann Dail, Gale Lawson, Susan Wiser, Charlie Griffin, and Jean Sorrell. The megaphone was one of several pieces of Civil Defense equipment carried on the fire department's Emergency Rescue Squad. (*RT* photograph.)

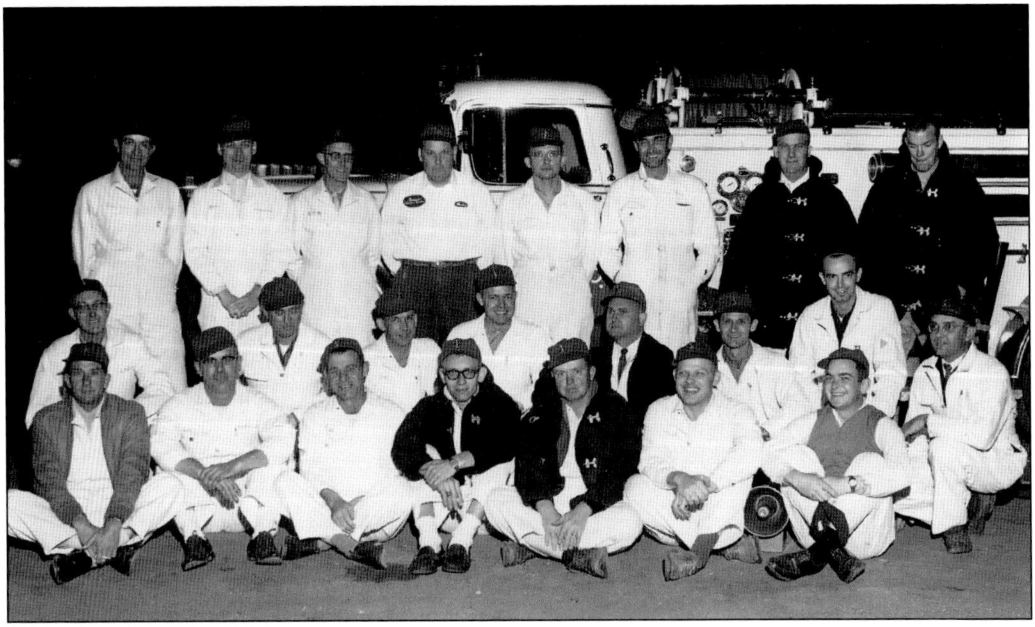

WESTERN BOULEVARD FIRE DEPARTMENT, 1958. Shown from left to right are (front row) Henry Garrison, Norman C. "Dick" Massengill, Biff Abbott, Skip Fox, Fred A. Burke, Robert Giddens, and C. Mitchell Godwin; (middle row) G. Rex Mann, F. Dale Graham, Ira O. Schaub Jr., James A. "Jimmy" Hewitt, Harold Morris, Edmond L. Johnson, Alex R. Russell Jr., and Bill Dupree; (back row) Charlie R. Weathers, Bobby Massengill, Bruce Brundage, Henry Wilder, unidentified, Mitchell Garbelly, Judson T. "Judd" Watkins, and unidentified. (Courtesy of Elaine Massengill Kurtz.)

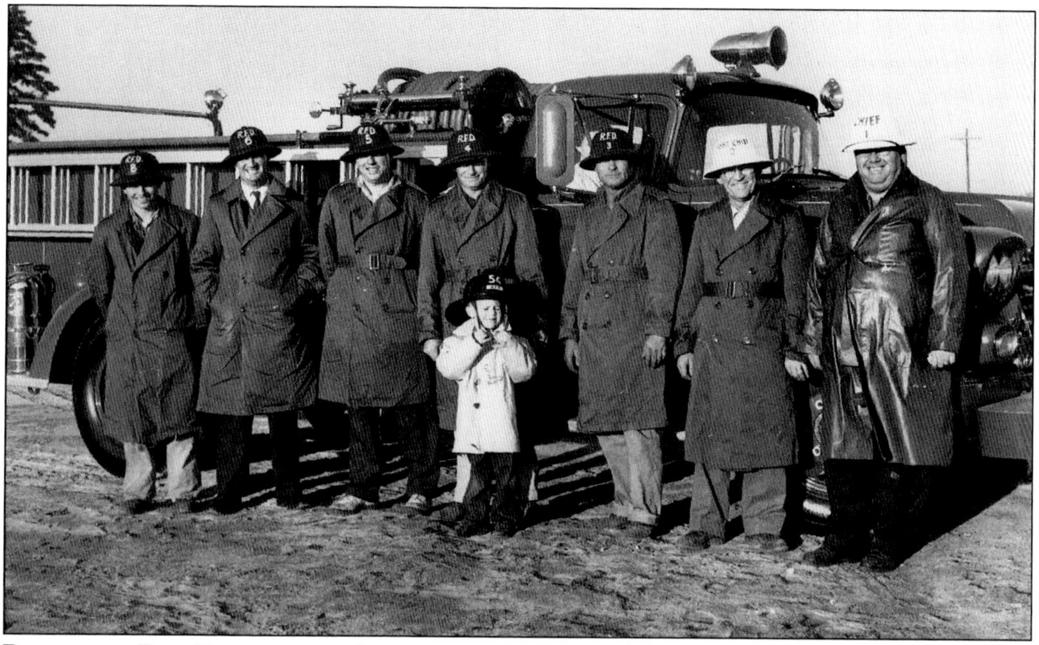

ROLESVILLE FIRE DEPARTMENT, JANUARY 1959. Pictured from left to right are Roy Ed Jones, June Privette, Curtis Underwood, Carl L. Scarboro, "mascot" Phil Perkinson, Wade Young, Asst. Chief Neal Rogers, and Chief H.E. Perry. The fire department was six months old at the time of this picture. (*RT* photograph by E.J. Parkins.)

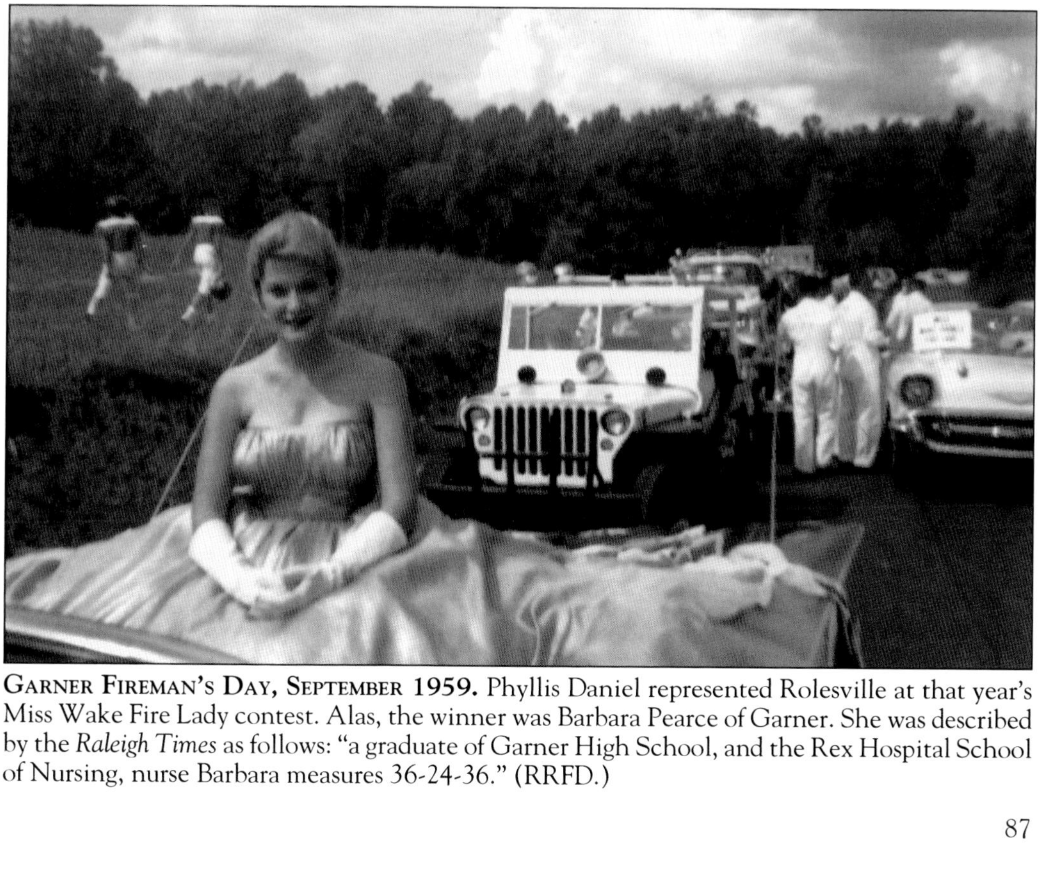

GARNER FIREMAN'S DAY, SEPTEMBER 1959. Phyllis Daniel represented Rolesville at that year's Miss Wake Fire Lady contest. Alas, the winner was Barbara Pearce of Garner. She was described by the *Raleigh Times* as follows: "a graduate of Garner High School, and the Rex Hospital School of Nursing, nurse Barbara measures 36-24-36." (RRFD.)

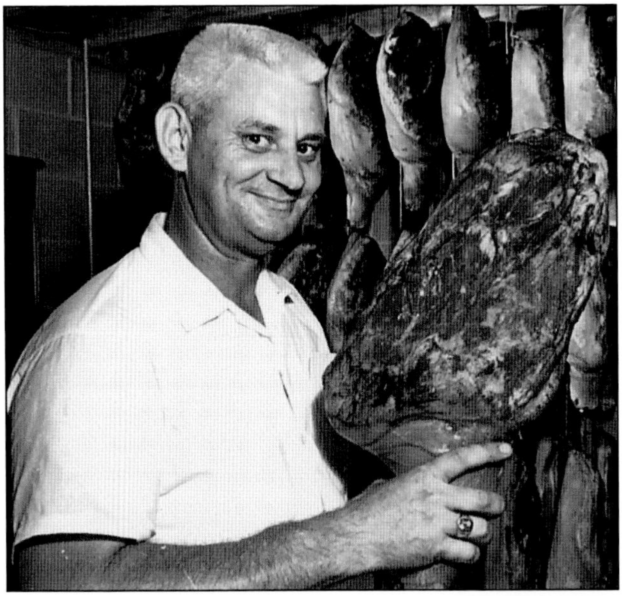

Luke Godwin, June 1960. Raleigh fire captain Lucius Q. "Luke" Godwin displays the hams he cures in his spare time. Though the location of his "salt room" was a secret, curing was done in a cinderblock building behind his house at 722 N. Bloodworth Street. Curtains kept the curing room dark, he noted in a newspaper article, a practice that resulted in "practically no trouble with flies." (*N&O* photograph.)

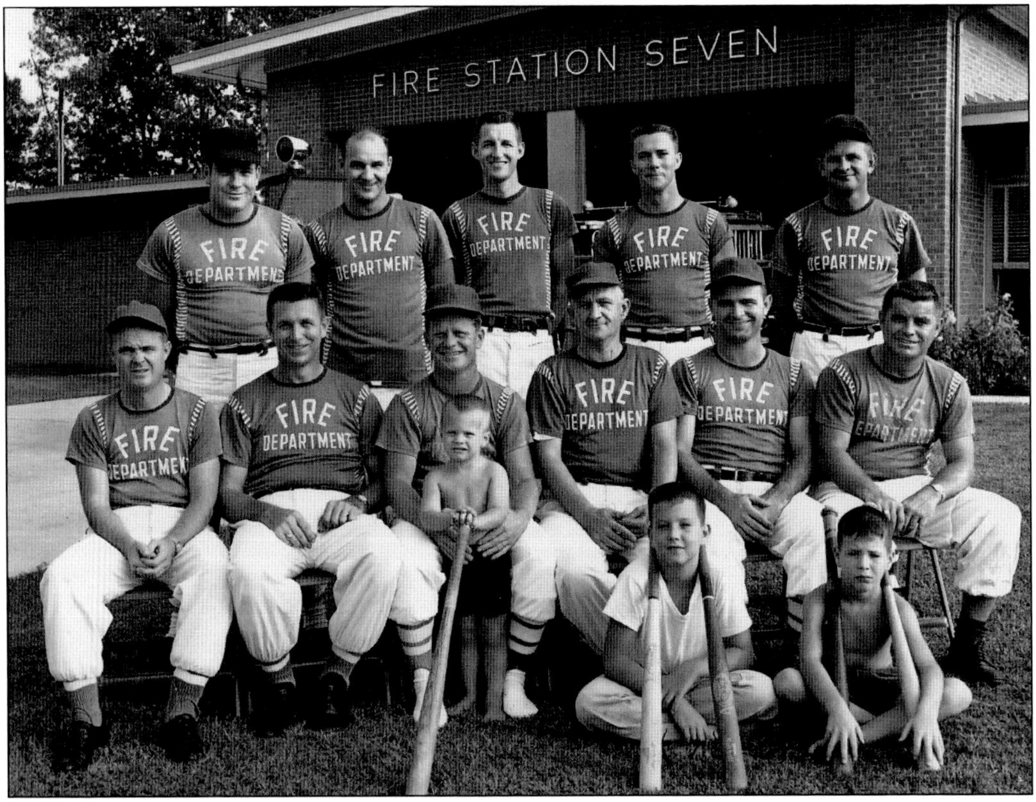

Champions, August 1960. Pictured in front of Raleigh Fire Station 7 are the winners of the AA Industrial-Municipal softball tournament. From left to right, they are (front row) Johnny Turner, C.T. May, Chester "Red" Stell, Harold Stephenson, Joe Partin, and Flentrel Blake; (back row) Henderson Taylor, Bobby Hicks, Charles Lloyd, Charles Narron, and Maylon Frazier. The batboys are Brit Stell, Terry May, and Randy May. (*RT* photograph by Ken Cooke.)

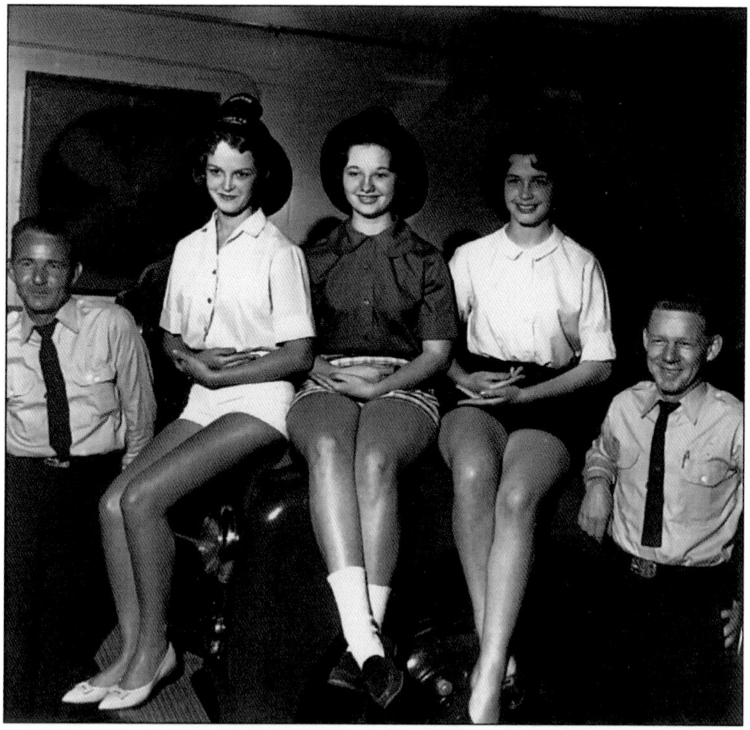

Honorary Members, May 8, 1961. Smiling with Jack Morris (left) and Calvin Beck (right) are three honorary members of the Cary Fire Department. From left to right, Sandra Cross, the reigning Miss Fire Department; Janet Bland; and Sandra Johnson would ride in the upcoming May 13 Fireman's Day parade. (*RT* photograph.)

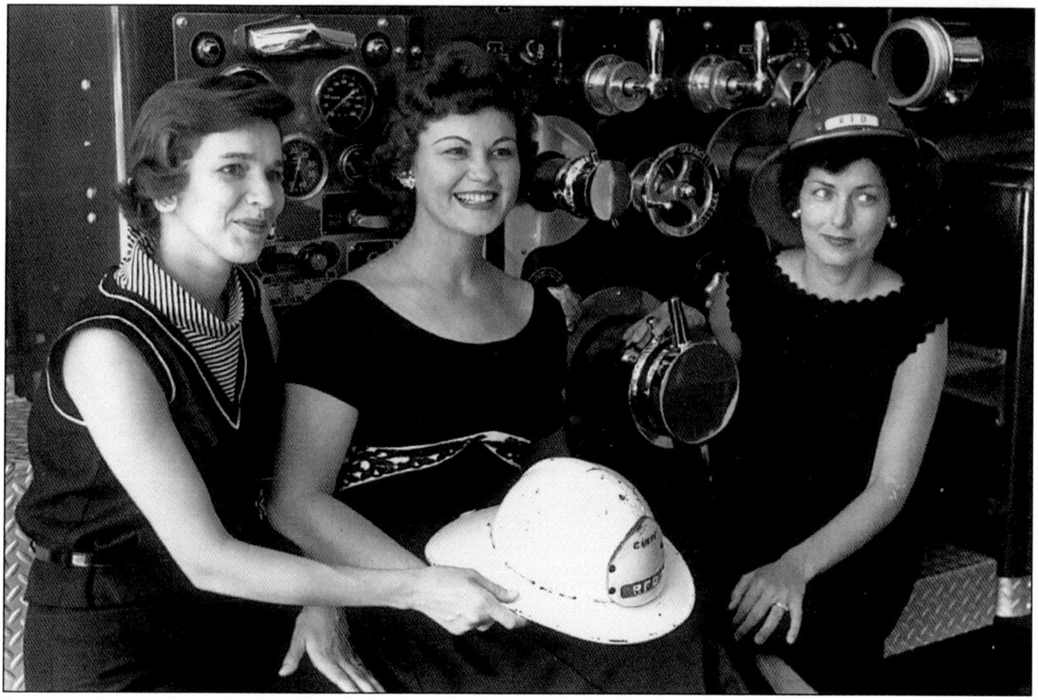

Ladies Auxiliary, September 1961. Pictured from left to right are Shirley Bowling, Peggy Hedrick, and Javain Partin. Formed by firefighter's wives in 1951, the Raleigh Fire Department Ladies Auxiliary performed charity work, sponsored social events, and served refreshments at fire scenes. (*RT* photograph.)

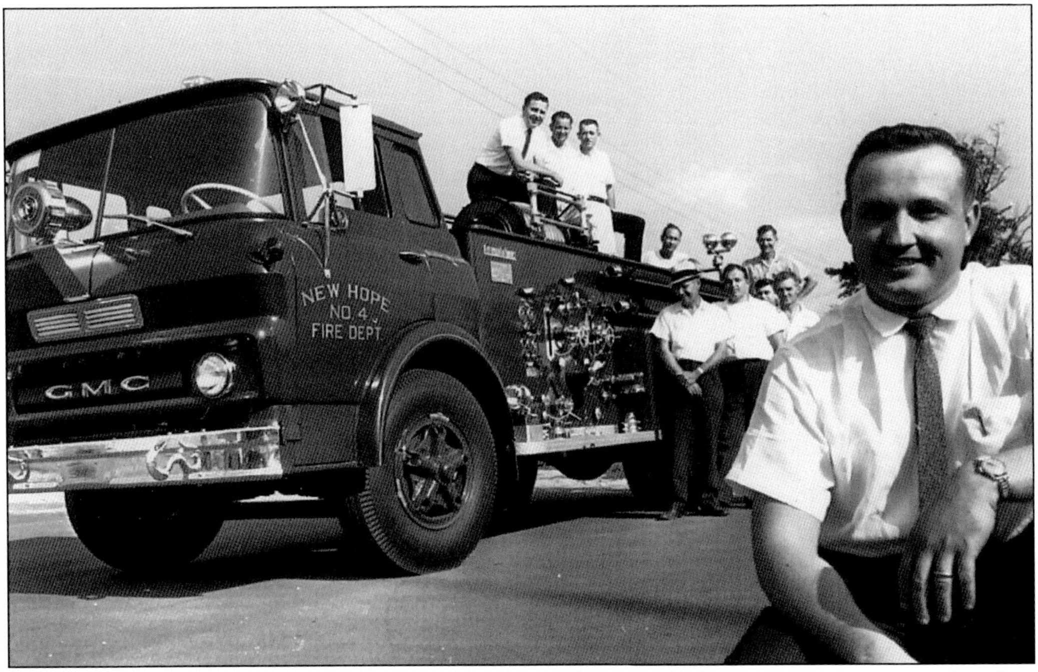

New Hope Fire Department, July 1962. Fire Chief G. E. Rosenberger poses with other members in front of their new 1962 GMC/American LaFrance fire engine. The 750-GPM, 500-gallon pumper cost $16,289. (*RT* photograph.)

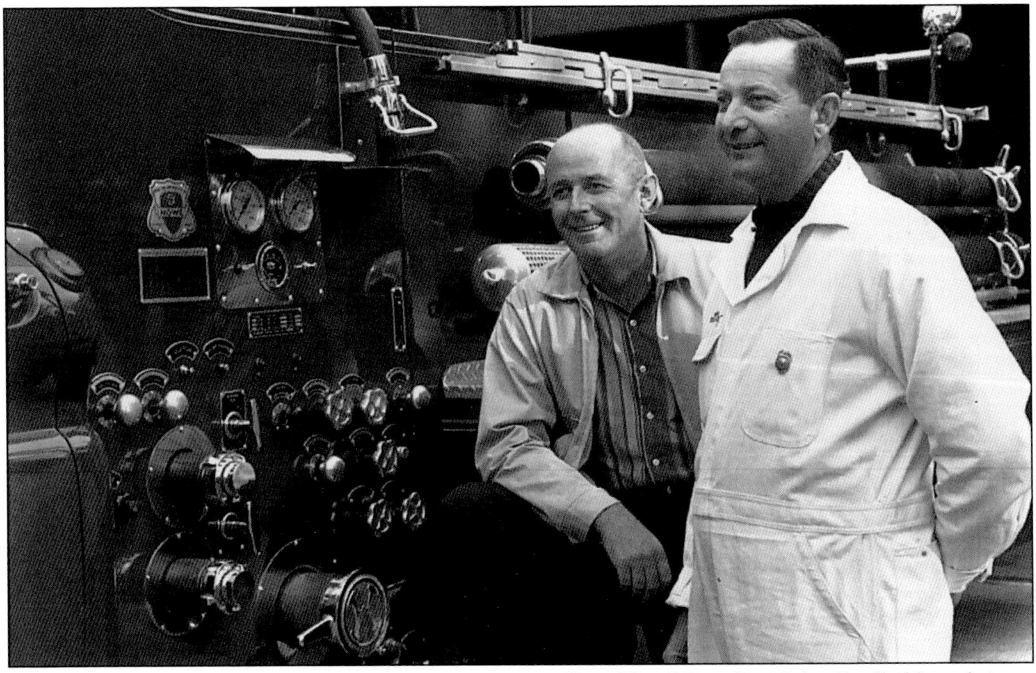

Swift Creek Fire Department, November 1962. Fire Chief Stanford Mizelle (left) and Asst. Chief Ransom W. Smith Jr. (right) are shown. Formed two years earlier, the fire department had 40 volunteer firefighters serving 630 families in the Swift Creek community southwest of Raleigh at the time of this picture. (*RT* photograph.)

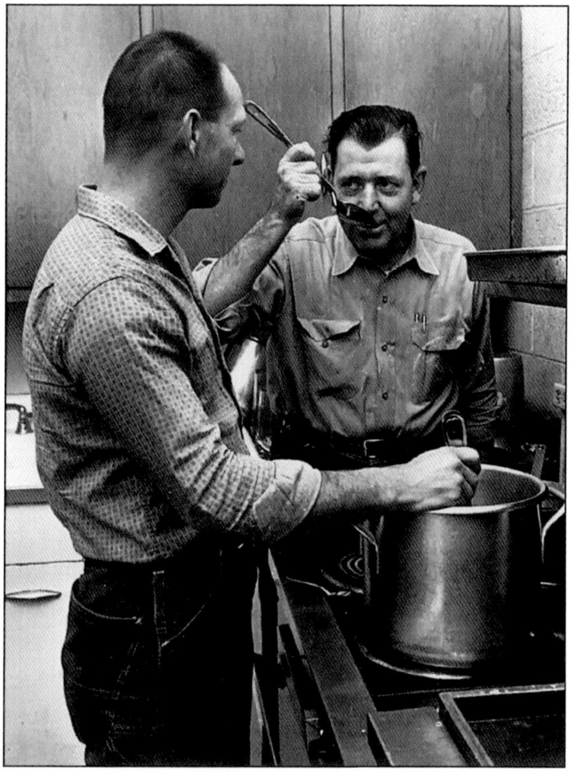

HENHOUSE HAVOC, JANUARY 1963. Raleigh firefighter Gene Alford (right) samples the chicken stew of Charles Woods (left) in the kitchen at Station 1. Woods' recipe: "Take five chickens, boil until you can pull all the meat off the bone. Cook the pastry separate, taking about 30 to 40 minutes (use plain flour). While pastry bakes, take one dozen boiled eggs and cut into chicken after bird is done. Add two sticks of melted margarine or butter, six spoonfuls of salt and one of pepper. After pastry is done, boil for 10-15 minutes until juice thickens." Yum. (*RT* photograph.)

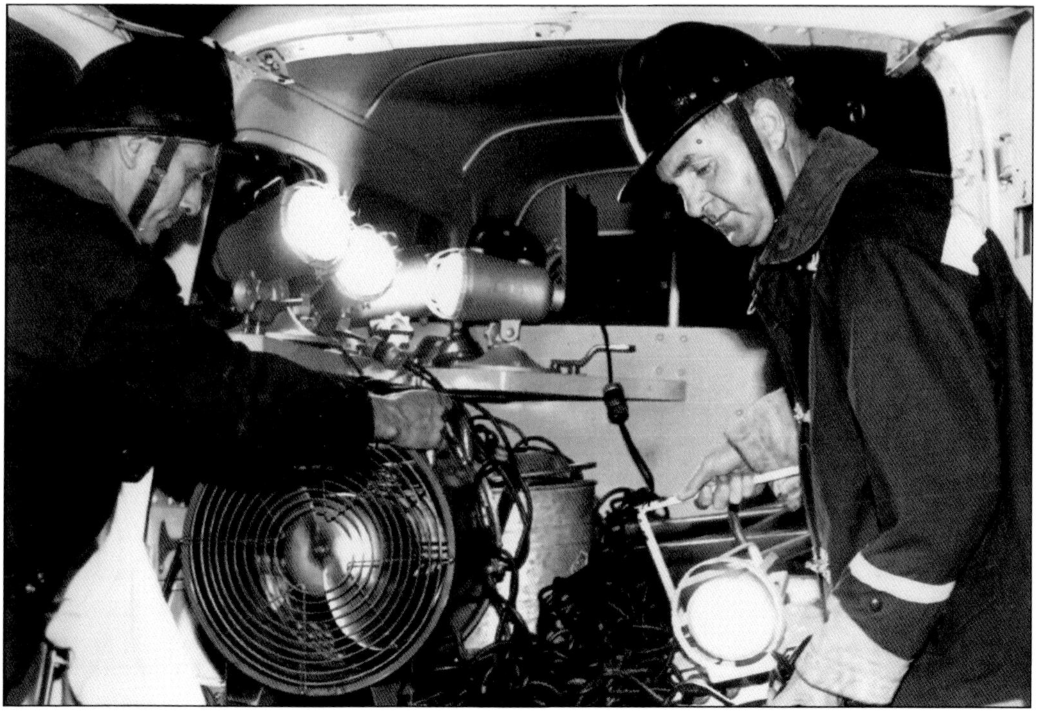

EMERGENCY LIGHTING, c. 1962. Garner firemen Joe Jackson (left) and Horace Barrow (right) unload smoke fans and spotlights from their 1952 GMC power unit. (GFD.)

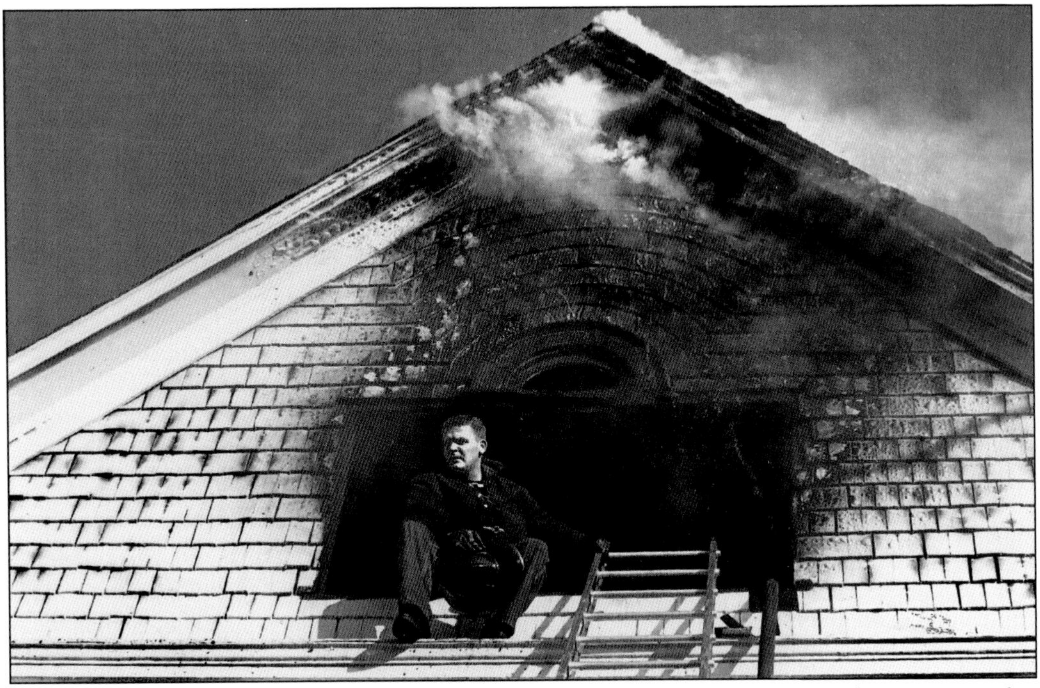

FRESH AIR, DECEMBER 21, 1963. Hubert Altman takes a break after extinguishing an attic fire at 10 Enterprise Street in Raleigh. (*N&O* photograph by Warren Uzzle.)

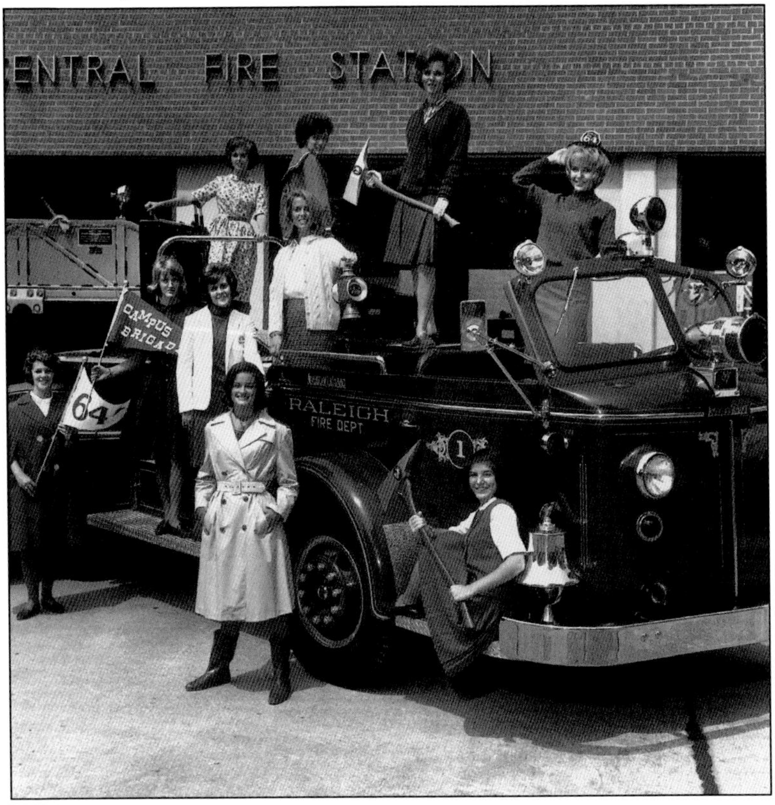

FASHION STATION 1, AUGUST 1964. Posed for a Hudson-Belk newspaper ad are models from Broughton, Cary, Enloe, Garner, and Millbrook high schools. Outfits include a "Yankee Peddlar's Green corduroy shirtdress" ($10.99), a "blue tweed Girltown double-breasted jumper" ($17.99), and an "Old Colony's Patriot Blue Blazer Sweater and Walking Skirt" ($16.99 and $14.99). (*RT* photograph.)

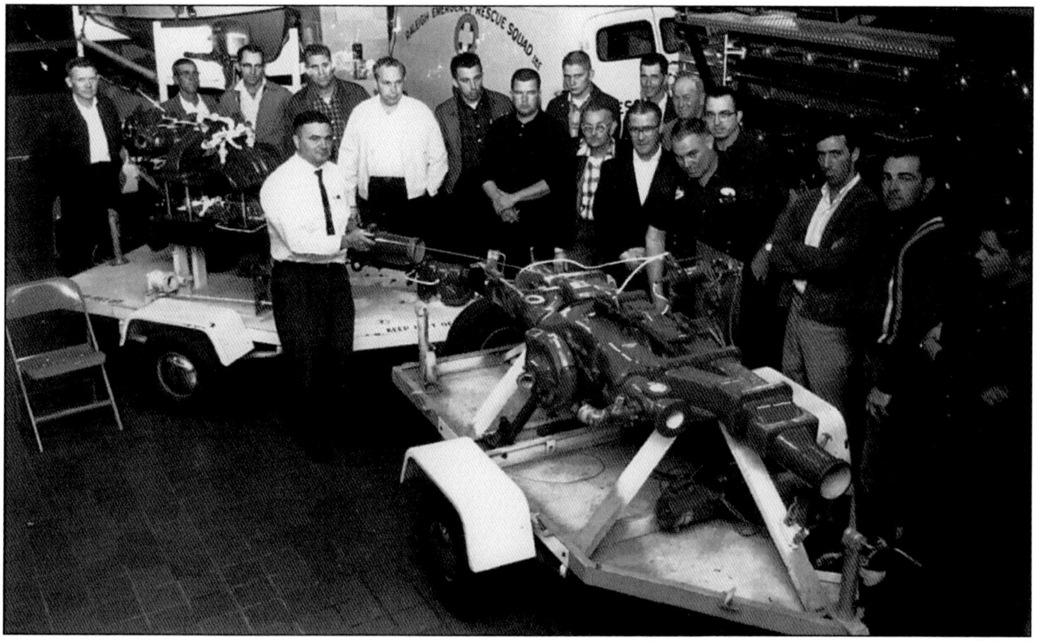

Volunteer Training, October 1965. Instructor David Lee demonstrates at Raleigh Fire Station 1 during an annual fire services school for Wake County volunteers. The four-day program included courses in pump operations, fire prevention practices, and volunteer fire department administration. (*N&O* photograph.)

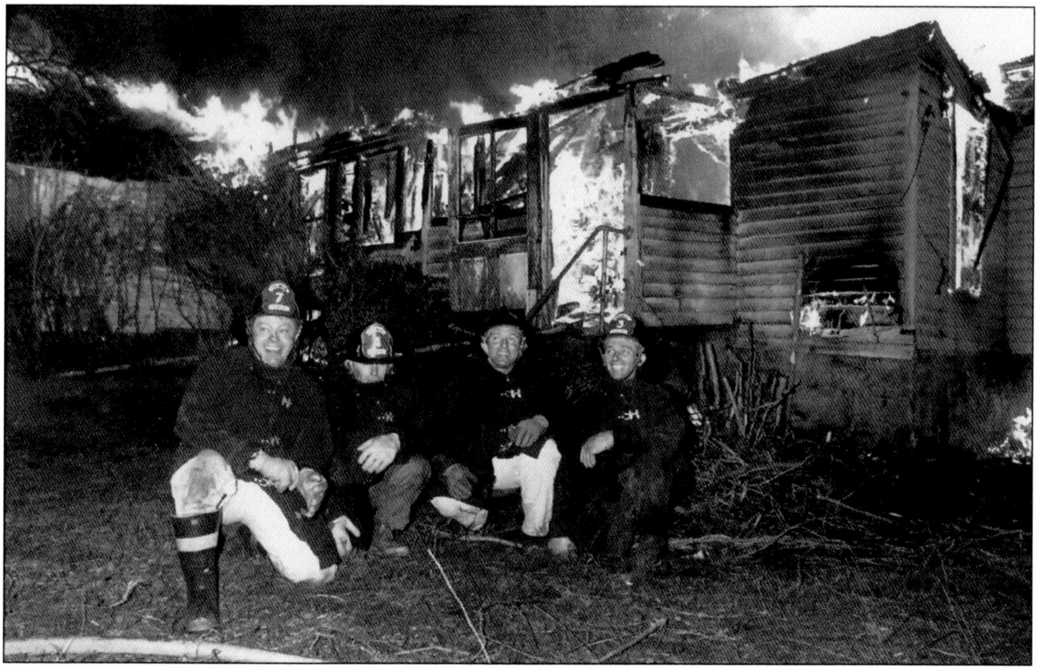

Durham Highway Fire Department, April 1966. Pictured at an evening exercise are, from left to right, Floyd Bailey, Tom Hamilton, Bill Hartness, and Tommy Ray. An abandoned house off Leesville Highway was burned as part of a 12-hour firefighting course sponsored by W.W. Holding Industrial Center. (*RT* photograph by Martin Rogers.)

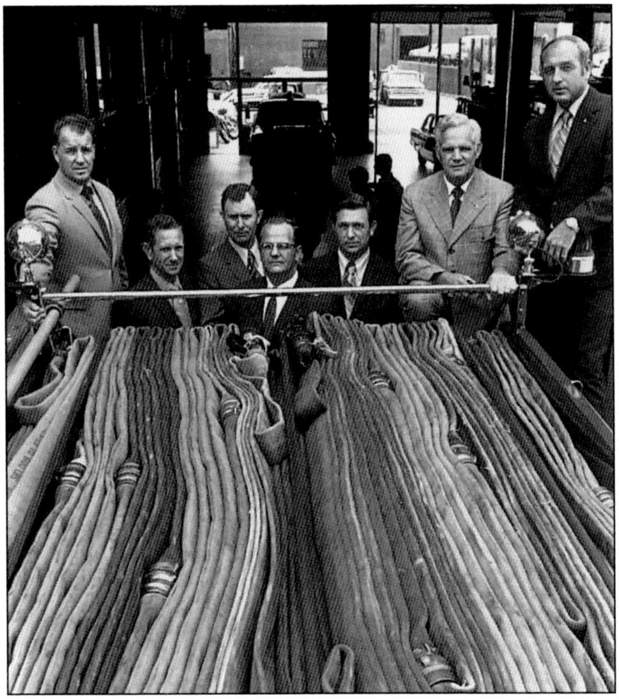

New Chiefs, August 1971. The rank of district chief was created after a massive reorganization of the Raleigh Fire Department administration. Shown from left to right are appointees Walter R. Mabrey, Norman W. Walker, Drewey H. Williams, William B. Hamilton, C.T. May, Stephen J. Talton, and Rufus E. Keith. The seven were selected from 19 candidates on the basis of written tests, interviews, and evaluations. Their salary range: $9,336 to $12,500. Their collective time in service: 145 years. (*RT* photograph by Harry Knickerbocker.)

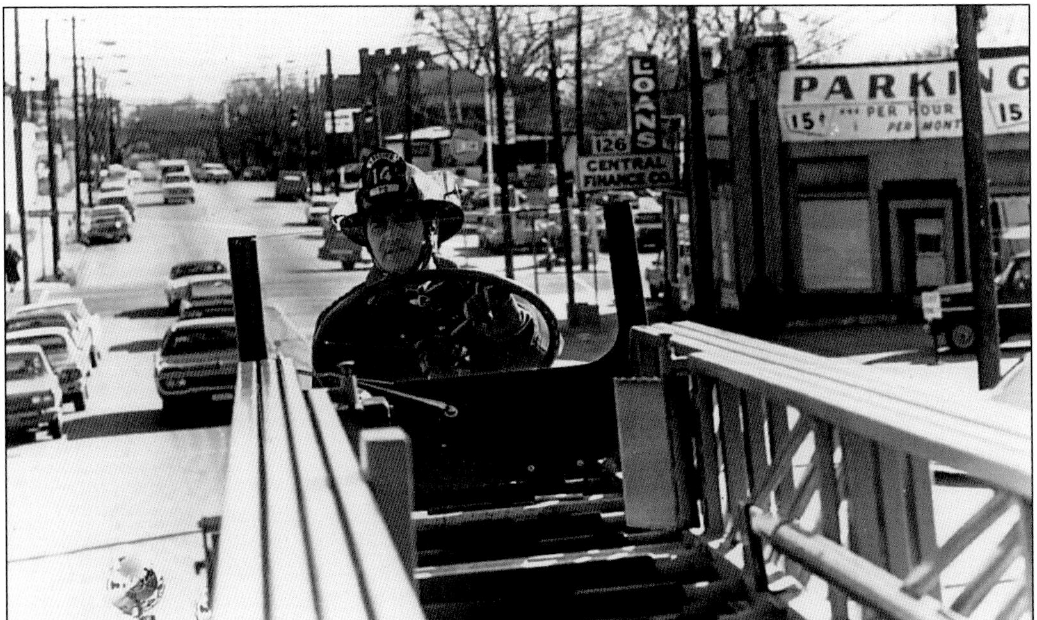

Tillerman, February 1972. Raleigh firefighter Charles Hinton steers the rear of Truck 1. (*N&O* photograph.)

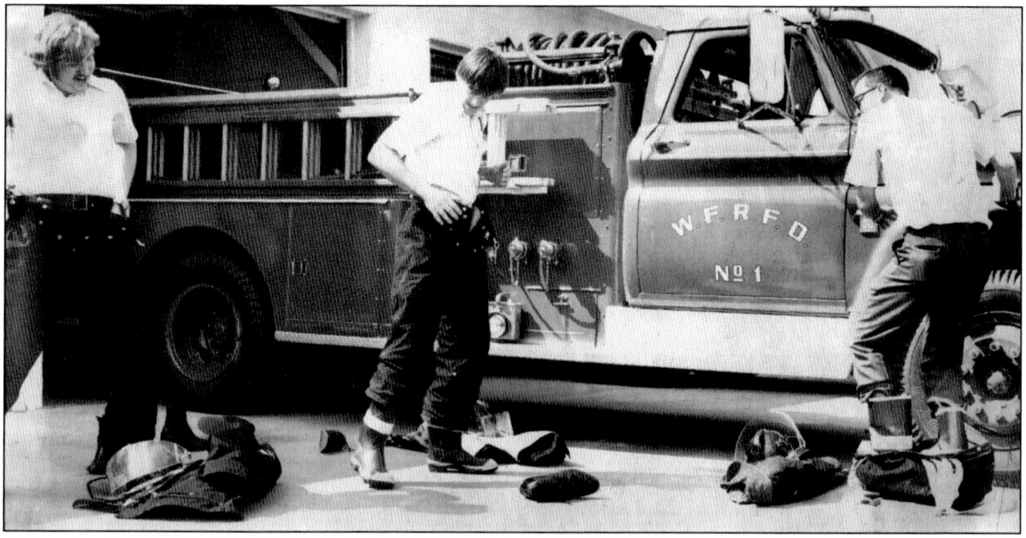

GETTING DRESSED, MAY 7, 1972. Pictured at a Wake Forest Rural Fire Department open house are, from left to right, Jimmy Keith, Randall Cooper, and A.C. Hall. (WFFD.)

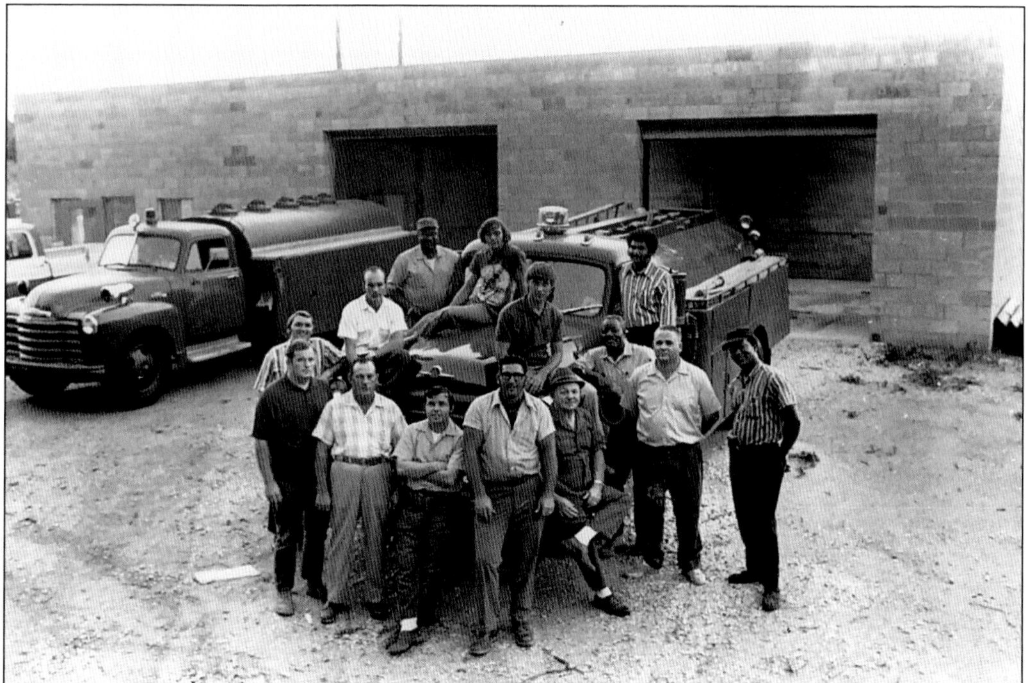

HOLLY SPRINGS FIRE DEPARTMENT, AUGUST 1972. Members of the newly operational fire department pose in front of their fire station on Holly Springs-Corinth Road. Built by the volunteers over a series of Saturdays, the cinderblock station first housed a 1953 GMC tanker (left) and a 1955 International tanker (right.) Fire Chief Jimmy Holland, front row on left, is pictured with Pete Brewer, Jerry Ellis, Floyd Whitaker, Jimmy Yarborough, Jim Russell, Billy Boone, Wick Holland, Ray Thaxton, Wardell Greene, Charles Whitaker, Larry Bryant, Henry Cotton, and Howard Witherspoon. (*RT* photograph by Warren Uzzle.)

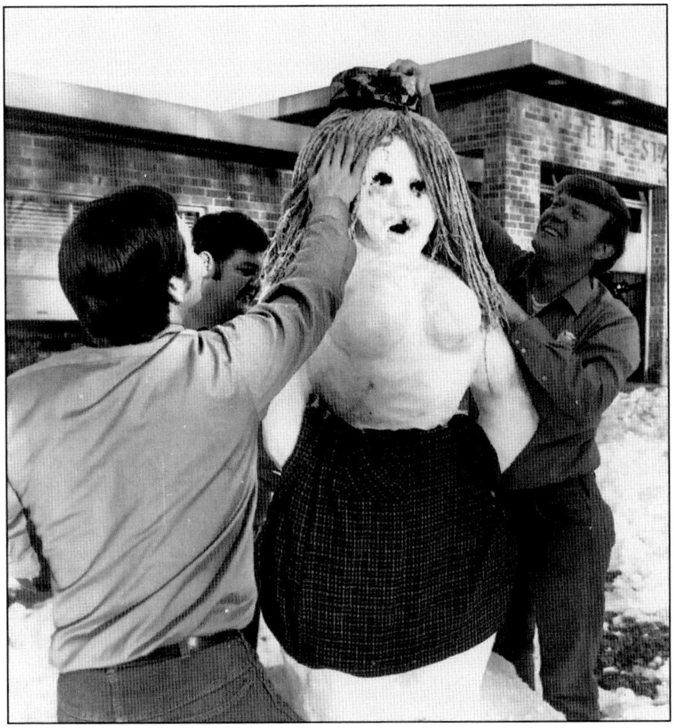

Snowwoman, January 11, 1973. After nearly seven inches of snow fell in the capital city, Jimmy Faison, Bobby Smith, and Glen Grice, pictured from left to right, decorated the lawn of Fire Station 5 with this anatomically correct display. (*N&O* photograph.)

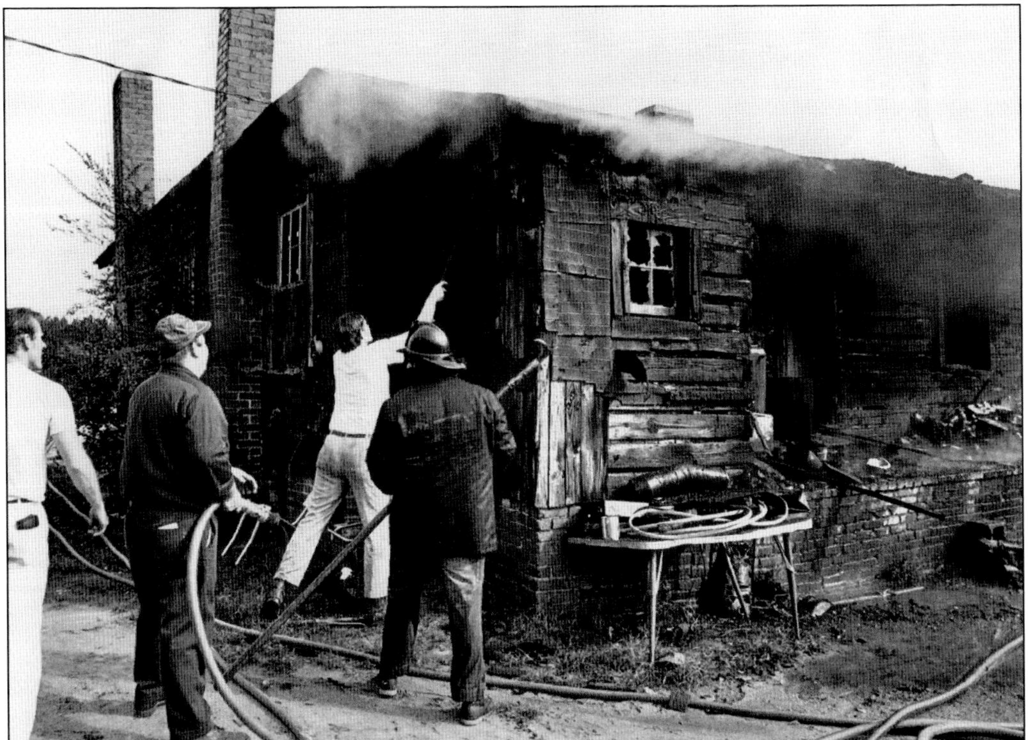

House Fire. Pictured at a house fire in Eagle Rock are Wendell firefighters Bob Jones (left) and Billy Ray Fuller (holding hose). (WFD.)

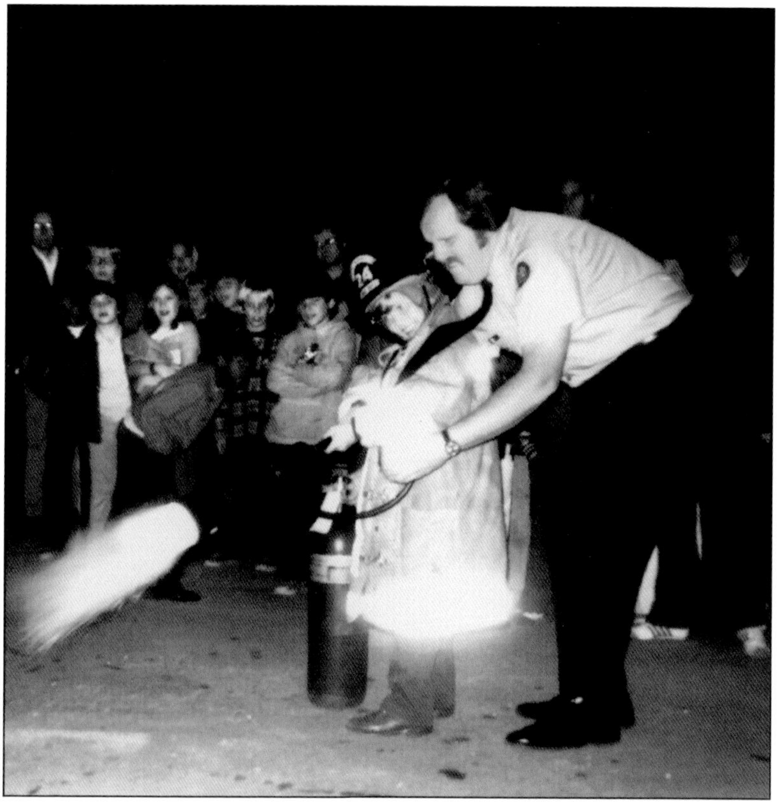

DEMONSTRATION, MAY 1974. Cary firefighter "Big John" Ruth is pictured at Fireman's Day. The 6-foot, 6-inch fireman was one of the town's 15 full-time firefighters. During the 1970s, the number of paid personnel increased from six in 1970 to 26 in 1980. The fire department also continued using some volunteers until the early 1980s. (CFD.)

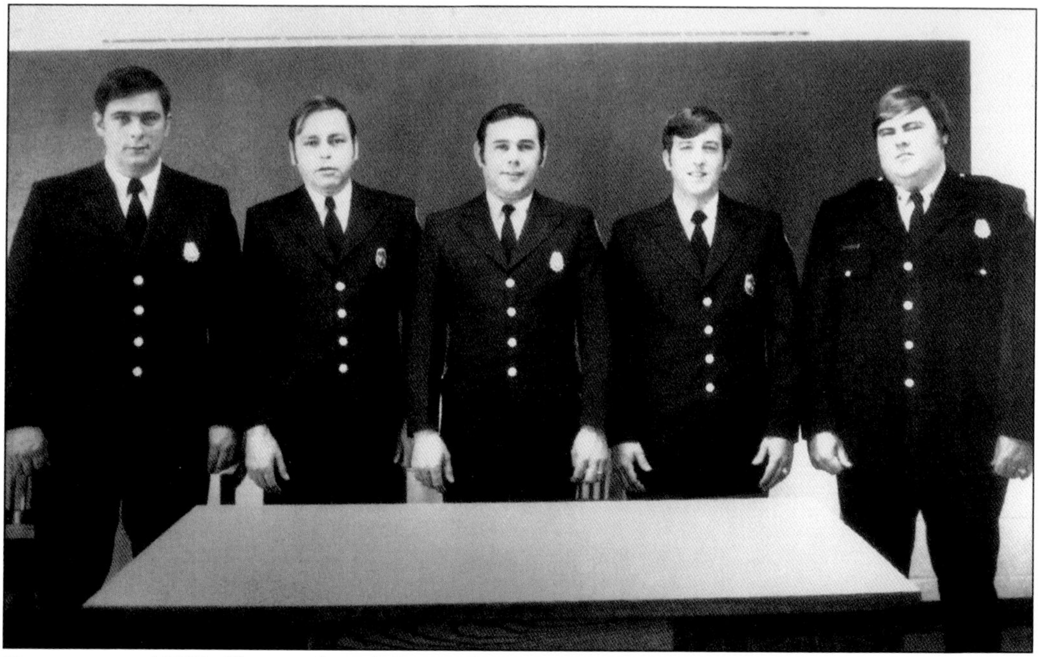

OFFICERS, C. 1975. Shown, from left to right, are Cary Fire Department officers Capt. Wayne House, Fire Marshall Wayne Frye, Fire Chief Terry Edmondson, Deputy Fire Marshall Don Daniels, and Capt. Dewey Poole. (CFD.)

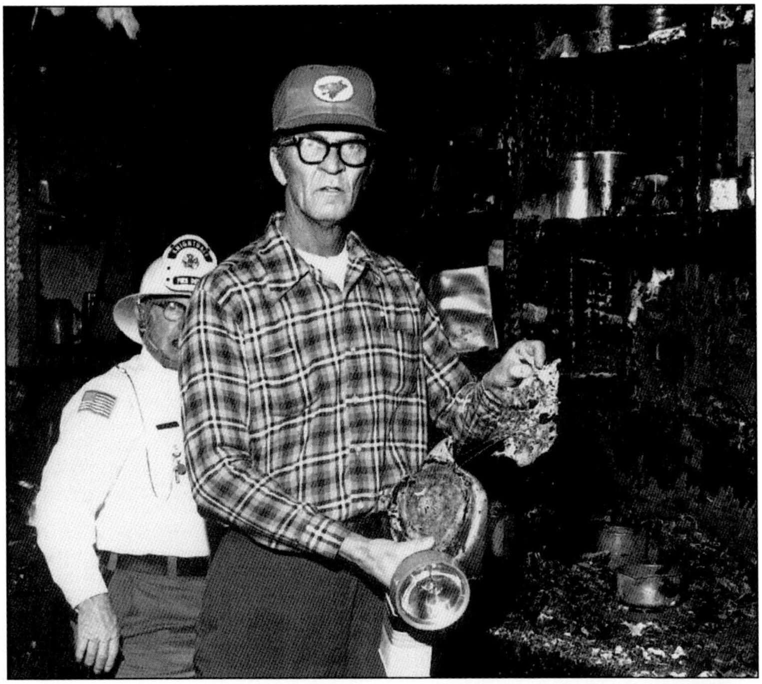

FIRE INVESTIGATION, JANUARY 23, 1976. Knightdale firefighter George Robertson holds a charred pan in a home on Route 2 near Wendell. Fire swept the ranch-style residence about 11:30 a.m., killing a 19-year-old male. About 35 firefighters and six trucks from Wendell and Knightdale responded. Also pictured is longtime Knightdale fire chief Bernice Wall. (*N&O* photograph by Lawrence Wofford.)

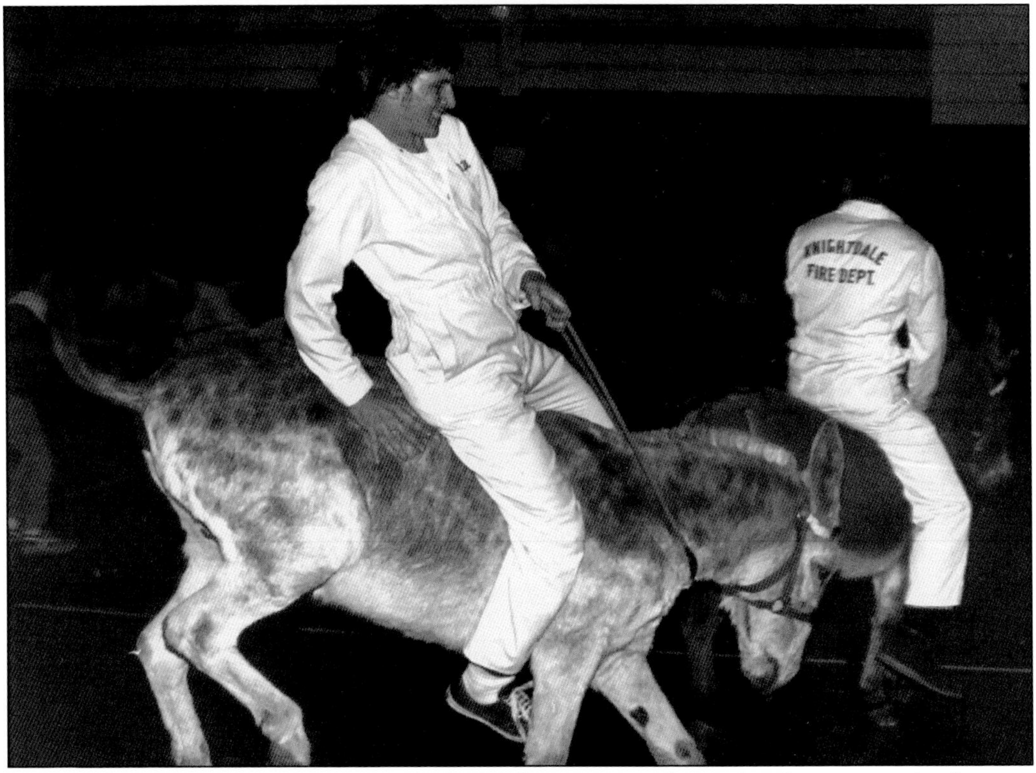

DONKEY BASKETBALL, MARCH 1977. Mike Miller is pictured at a fire department fund-raiser, astride an ass equipped with rubber shoes to protect the floor of the Knightdale Elementary School gym. (*GLF* photograph.)

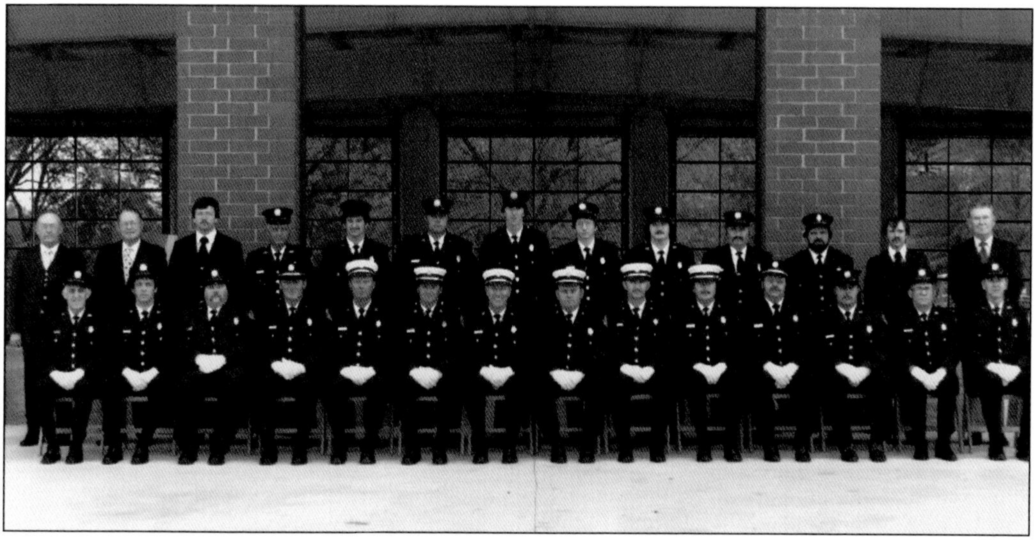

FUQUAY-VARINA FIRE DEPARTMENT, DECEMBER 1978. Pictured in front of their new fire station at 301 S. Fuquay Avenue are, from left to right, (front row) Walter Howard, Doug Knott, 2nd Lt. Phillip Bowden, 1st Lt. Ricky Stuart, Capt. T.C. O'Connell, Asst. Chief James Mauldin, Fire Chief Ed Schmelzer, Asst. Chief Johnny Jones, Charles Sauls, Capt. Mike Seawell, 1st Lt. William Dickens, 2nd Lt. Stuart Hair, Tommy Pleasant, and J.W. Mangum; (back row) Milton Lanier, William Kesler, David Markle, Jimmy Baker, Randy Egsegian, Jasper Weathers, David Mangum, Tim Matthews, Ed Gore, John Arnold, Tony Young, Fred Bass, and Gene Bradley. (FVFD.)

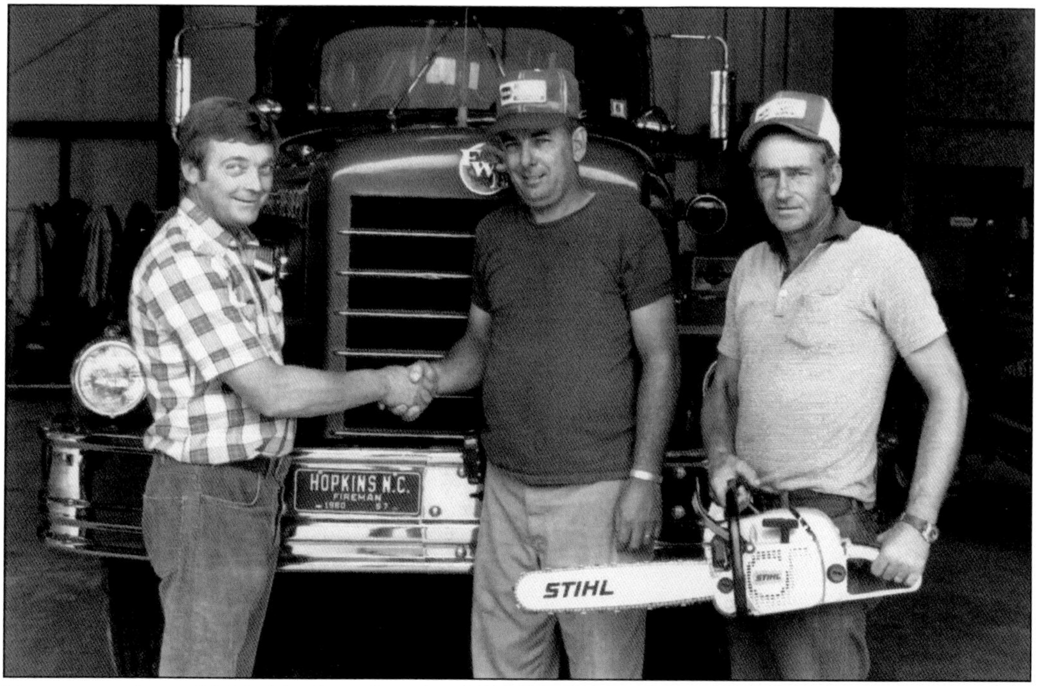

DONATION, JULY 1980. Hopkins fire chief Charles Hocutt (center) and assistant chief Donald Jackson (right) accept a new Stihl chain saw from Billy Adams (left) of Knightdale Tractor and Equipment. (*GLF* photograph by David Roberson.)

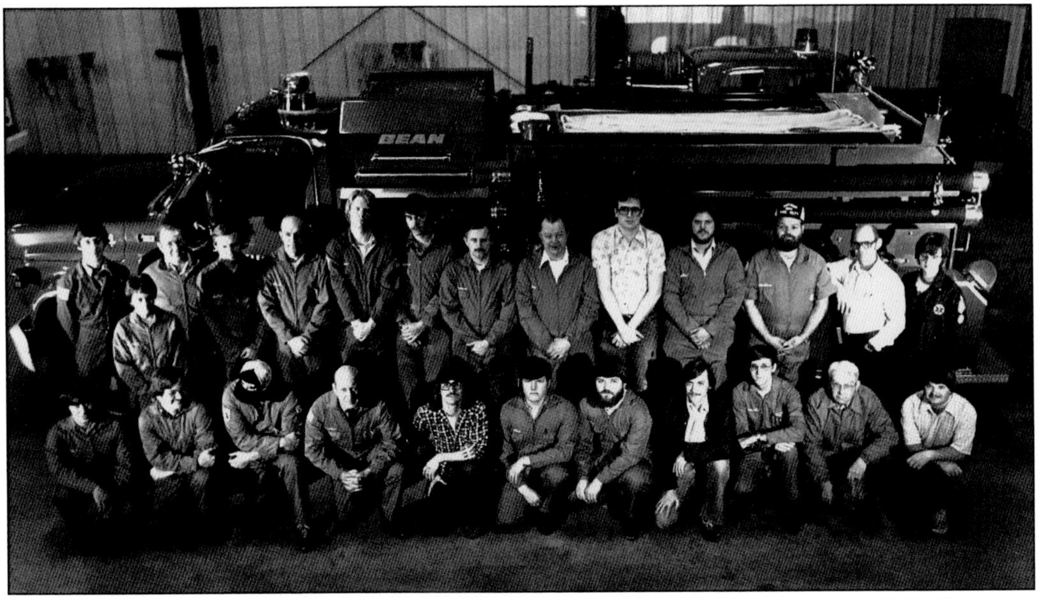

DURHAM HIGHWAY FIRE DEPARTMENT, C. 1980. Pictured inside Station 2 at 8312 Pinecrest Road are, from left to right, (front row) Ricky Ferebee, Mike Boyce, William Key "Bill" Sansone, Clyde Arvin, David Blackmon, James W. "Jim" McKay, Eddie Morton, Rex Craig, Einar Betancourt, Bill Davis, and Floyd "Sonny" Hester; (back row) Mike Abrams, Sheila Boyce, William A. "Bill" Fredette, Ernest O. Goodwin, John Green, Barry Cross, Nick Slobodzian, Charles Holleman, Floyd Bailey, Robin Laskey, Greg Johnston, John Muster, Bobby Layton, and Jan Holleman. (DHFD.)

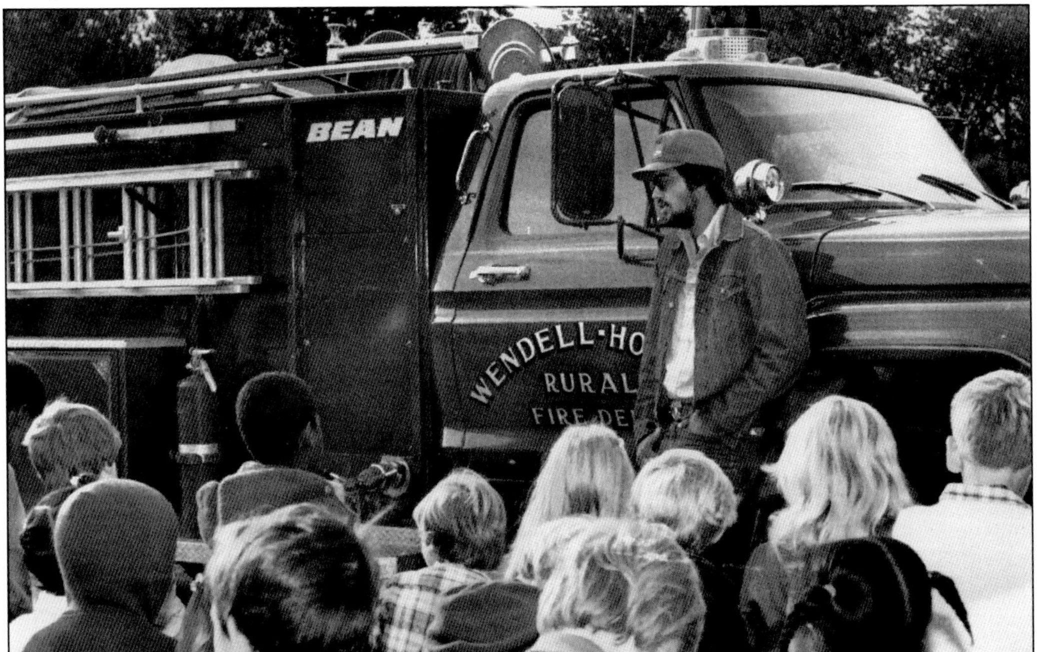

CARVER ELEMENTARY, OCTOBER 1981. Randy Raper shows a fire truck to first-graders at Carver Elementary School in Wendell. (*GLF* photograph by Saundra Freeman.)

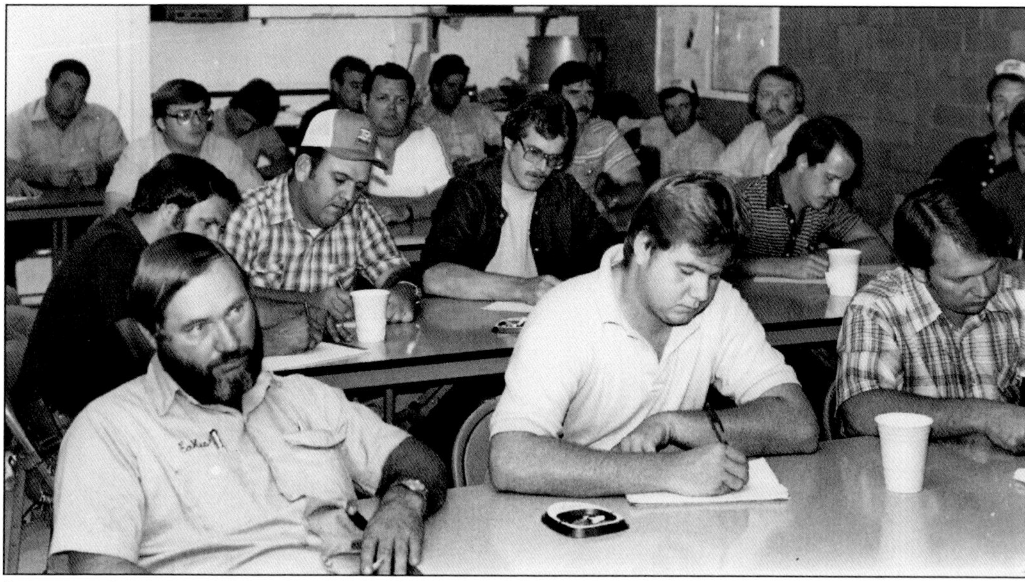

SCHOOL, SEPTEMBER 1982. Pictured at part of a nine-hour course on ventilation techniques from left to right, are Zebulon firefighters (first row) E.G. Eakes, Chris Ward, and Billy Mabe; (second row) Charles Merritt, Franklin Eddins, Gerald Lanier, David Gay, and Wilson Eddins; (third row) Milton Bryant Jr., Roger Strickland, Wayne Massengill, and Dickey Bryant; (fourth row) Willis Rogers, Kimbry Boykin, Danny Fogleman, and Waddell Mitchell. (*Zebulon Record* photograph by Edie Evans.)

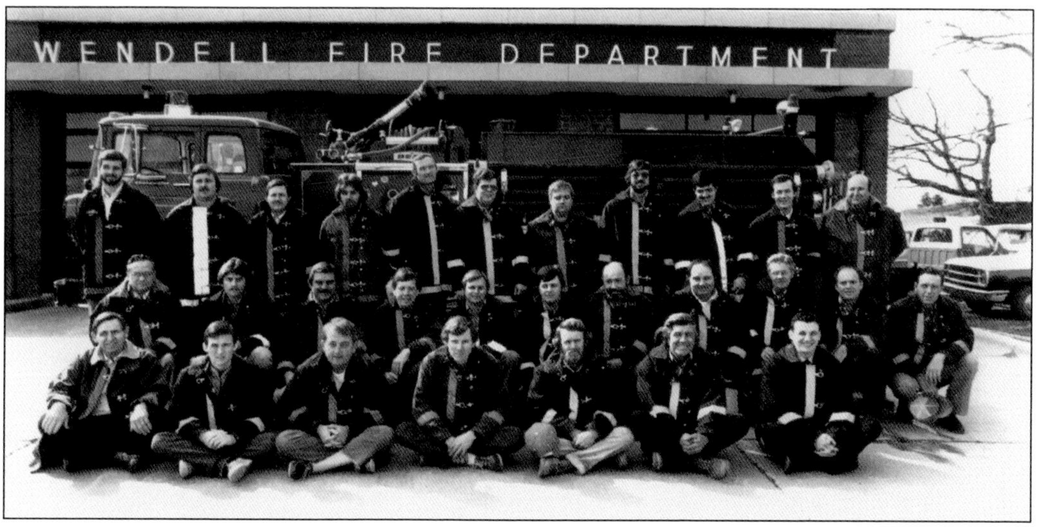

WENDELL FIRE DEPARTMENT, APRIL 1983. Shown, from left to right, are (front row) Nevell Davis, William Myron "Boody" Yeager, Tommy Coley, Cee Todd, Thomas Hinnant, Billy Raper, and Buddy Scarboro; (middle row) O.K. Strickland, Dean Ayscue, Ronald Barham, Mike Coley, Don Batten, Cliff "Tadpole" Richardson, Don Price, Fire Chief Herb Ramsey, Sam Anderson, Don Brady, and Marvin Nowell; (back row) John Underhill, Barry Allen, Sherwood "Weed" Adams, Terry Sasser, Thomas Strickland, Calvin Richardson, Johnny Keith, Randy Raper, Leon Cobb, William Ray Fuller, and Billy Ray Fuller. They're pictured at the Municipal Building on E. Fourth Street. From 1963 to 1984, the fire department occupied a one-story section in the rear. (*GLF* photograph.)

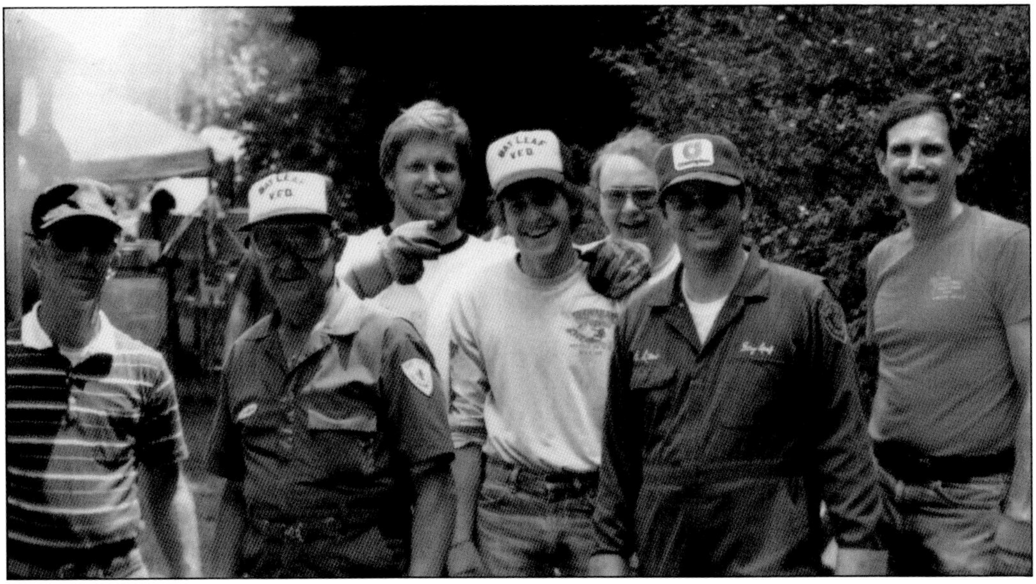

Barbecue, May 1984. Pictured at a Bay Leaf Fire Department's annual "Chicken Cook" are, from left to right, R.L. Harris, Lonnie Dean, Don Heres, David Rogers, Alton Ledford, Ersell Liles, and Mark Peterson. The popular event was discontinued at the end of the decade due to dripping fat. When the Falls Lake watershed was created, environmental restrictions prohibited the free release of grease, and the cost of containment was greater than the revenue generated by the fund-raisers. (BLFD.)

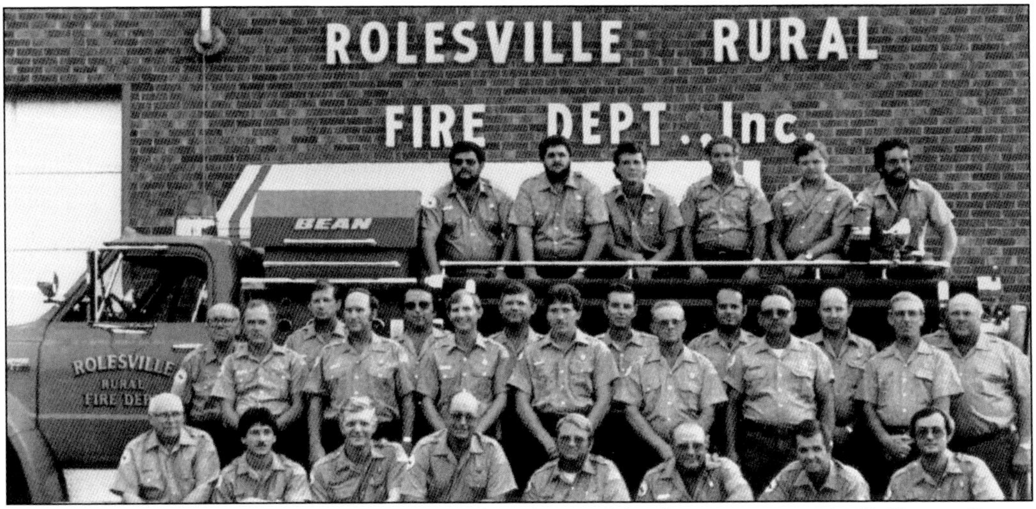

Rolesville Fire Department, 1984. Posed in front of the fire station at 104 E. Young Street are, from left to right, (front row) Louis Pearce, Glen Jones, Luther Perry, Bill Mitchell, Roy E. Jones, Duke Perry, Jimmy Gay, and Rodney Privette; (back row) H.E. Perry, Doug Jones, Ben Bartholomew, Curtis Mitchell, Doug Pearce, Donald Jones, Tommy Murray, Charles Jones, Jesse Guy, Kenneth Pearce, Douglas Young, Frank Pearce Jr., Larry Averette, Neal Mitchell, and Dennis Perry; (on truck) John Hugh Davis, Tommy Gay, Al Privette, Mike Mitchell, Bert Richards, and Craig Woodlief. The photograph appeared in the 1984 Rolesville Liness Club's third annual Community Birthday Calendar and was "compliments of Sarah's Hair Styling." (RRFD.)

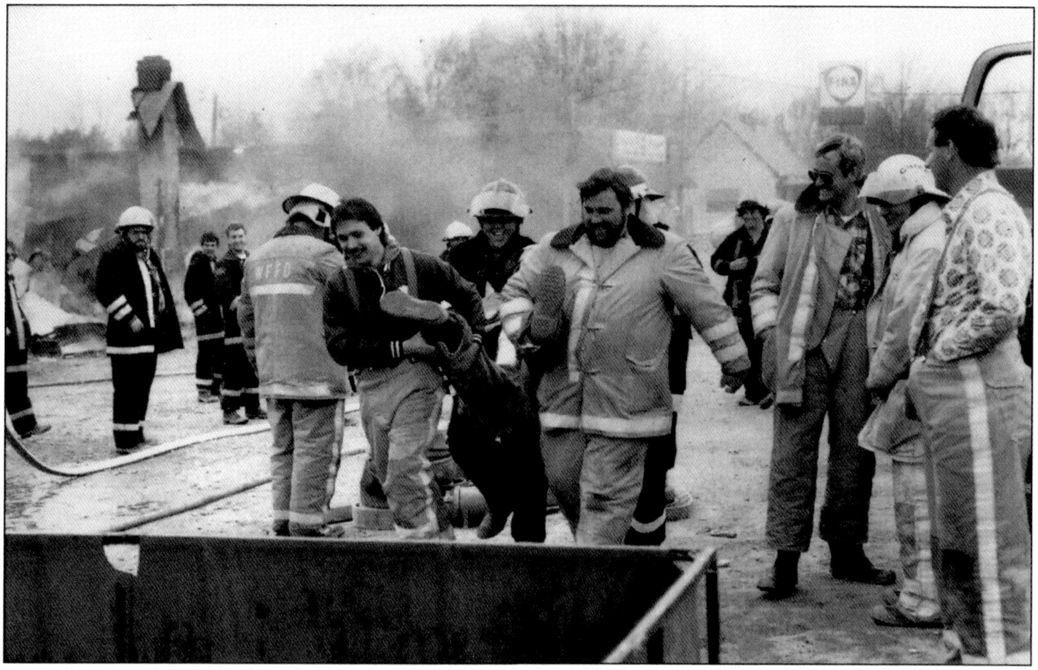

Horseplay, April 5, 1987. Lee Allred gets dunked in a drop tank at a Sunday afternoon exercise in Wake Forest. Sixty-four firefighters from nearly every department in the county attended the live burn on N. Main Street. The training event concluded a 12-hour course for rookie firefighters. North Carolina Department of Community Colleges Fire Coordinator Calvin Beck conducted the training. Wake Forest Fire Department Ladies Auxiliary members provided refreshments. Baths were free. (*WW* photographs by Greg Allen.)

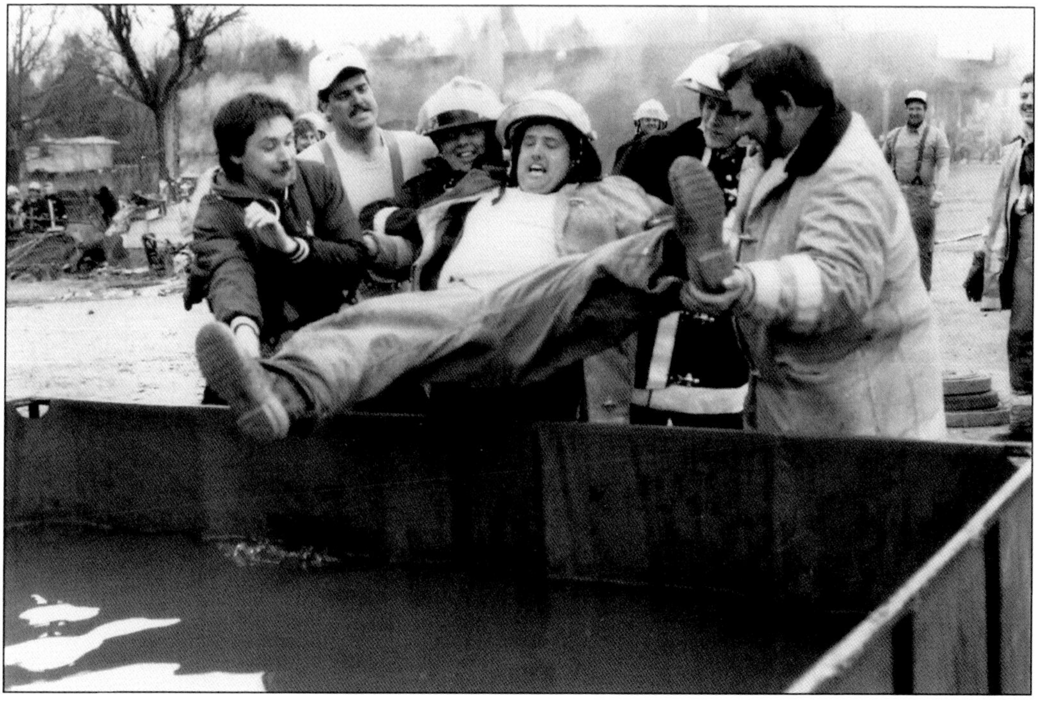

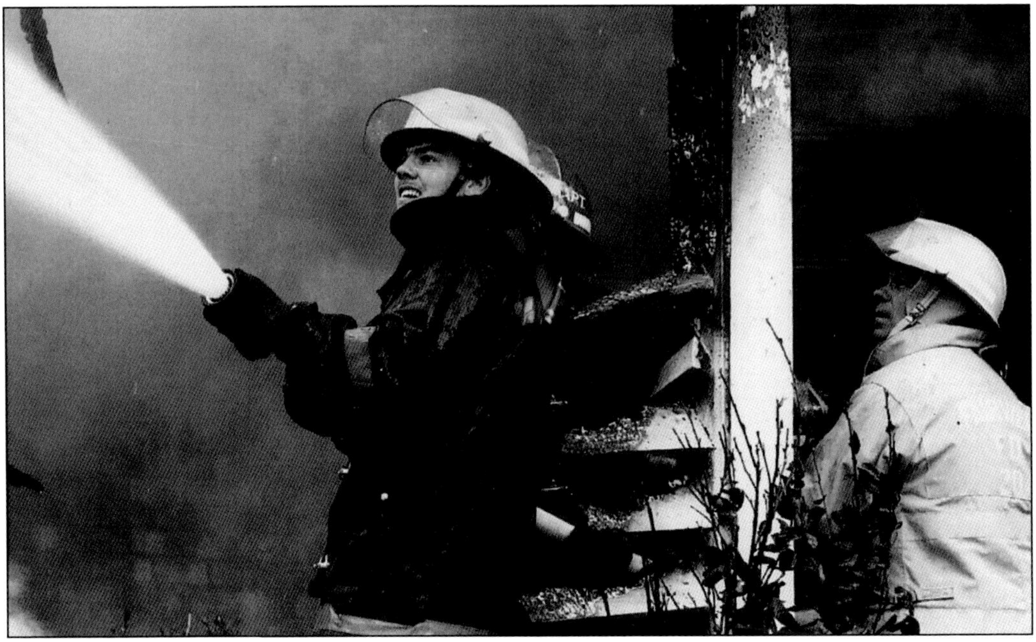

MUTUAL AID, DECEMBER 1, 1986. Fuquay-Varina firefighter Jonathan Randall (left) and Fairview firefighter Walter Ray Franks (right) are pictured at the home of Agnes Duncan on Rhamkatte Road. Units from Holly Springs, Fairview, and Fuquay-Varina responded to the 3 p.m. fire that started in a carport. (*Cary News* photograph by Robert Thomason.)

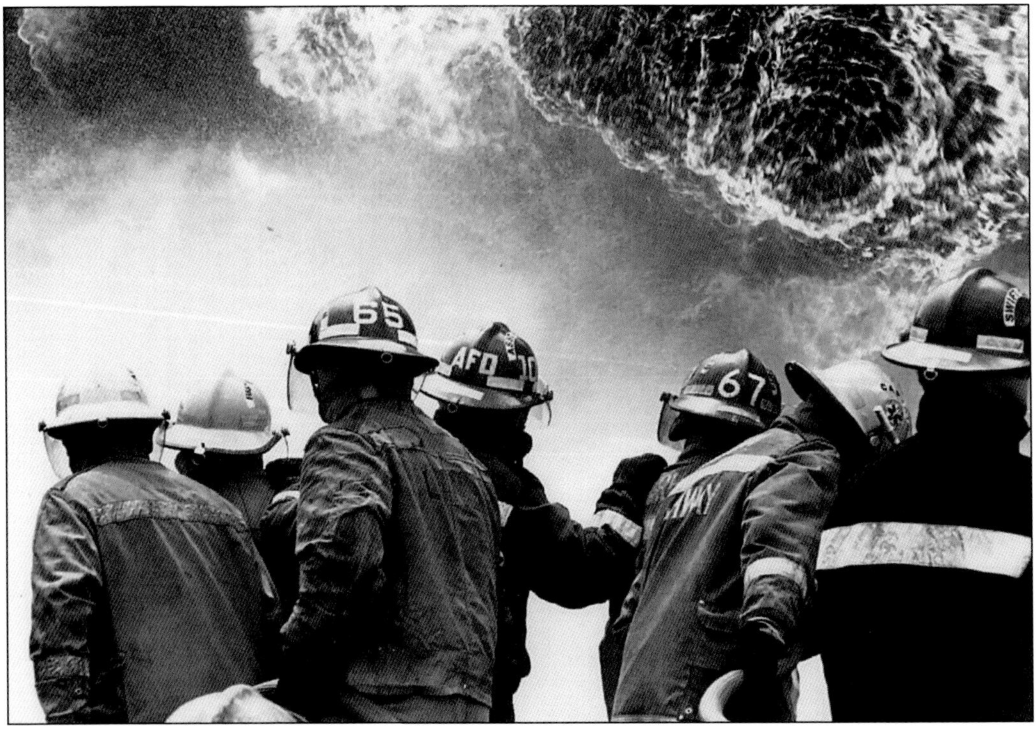

WALL OF FLAME, JUNE 1989. Volunteer firefighters practice at the Wake County Fire Training Center. (*N&O* photograph by Lance Powell.)

Five

Other Events

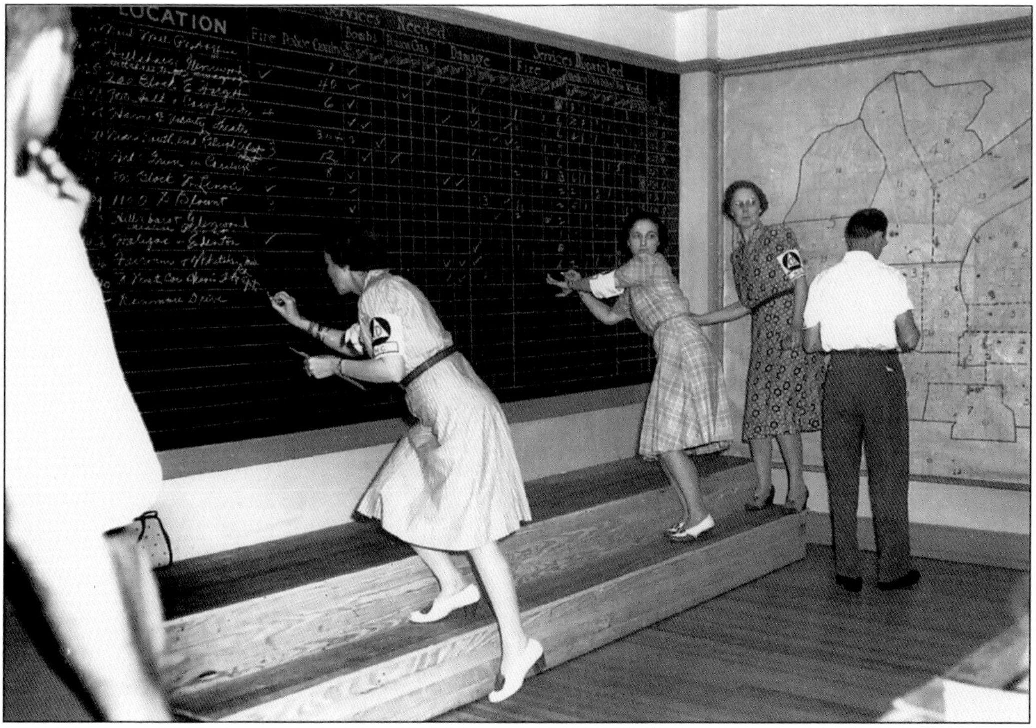

Air-Raid Practice, June 4, 1942. This Civil Defense control center came alive during Raleigh's first daylight air-raid practice. Factory whistles initiated the alert at 6:41 p.m., and the all-clear signal was sounded at 7:14 p.m. During the 33-minute period, the city suffered 76 mock casualties, including 20 people "injured" by the mock bombing of four houses on N. Person Street. Other communities around the state also conducted drills that day. In Burlington, several theater-goers fainted upon hearing the alarm. (*N&O* photograph.)

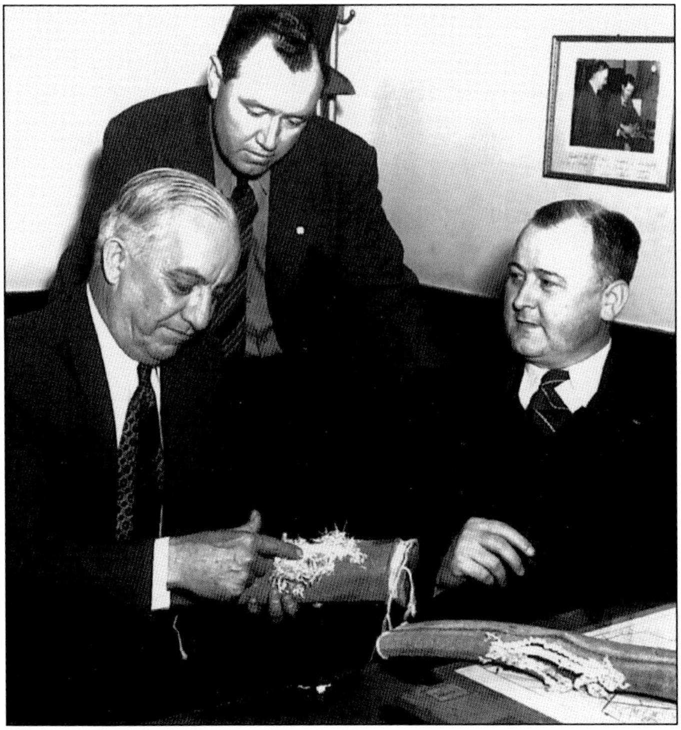

SABOTAGE, APRIL 1941. International Association of Firefighters President Fred Baer (left) points to a section of hose tested with acid that refuted claims of sabotage at Raleigh Fire Station 1. He found that damage to some 400 feet of fire hose was accidental. He also cited natural or accidental causes for both the alleged tampering of an eyewash solution and opened oil and water petcocks on one of the newest fire engines. Capt. Kenneth J. Smith (center) and Commissioner of Public Safety Robert C. Powell (right) are also pictured. (*N&O* photograph.)

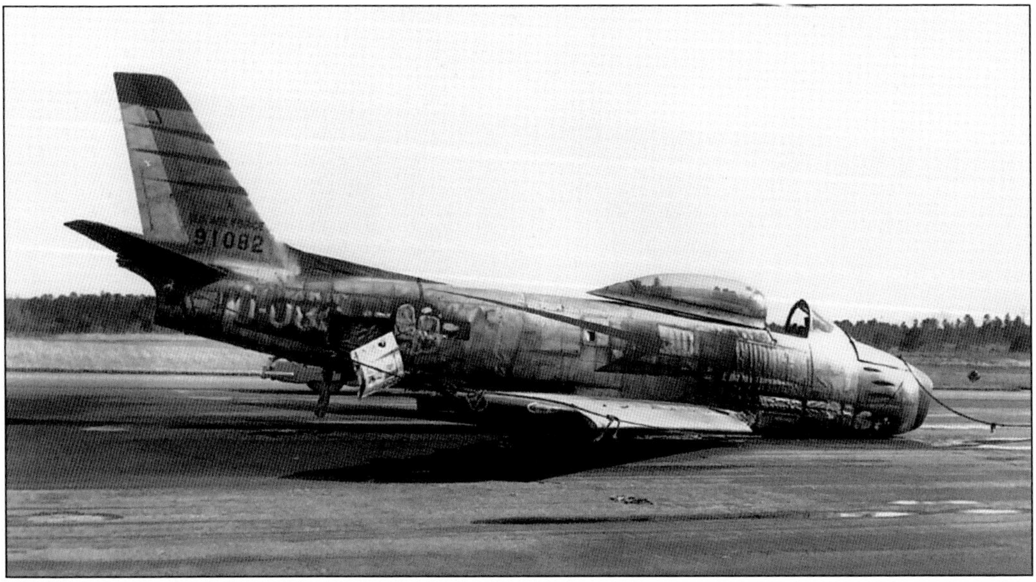

FIERY LANDING, APRIL 1, 1951. This Air Force Sabre fighter, one of four flying from Langley Field, Virginia, to Maxwell Field, Alabama, burst into flames after crash-landing at Raleigh-Durham Airport. Skidding almost 4,500 feet along the runway, 2nd Lt. Coy A. Austin leaped uninjured from the burning craft. His wheels-up landing ignited the wing-tip fuel tanks. Crash crews saved the jet from a total loss. (*N&O* photograph by R. W. Stephens.)

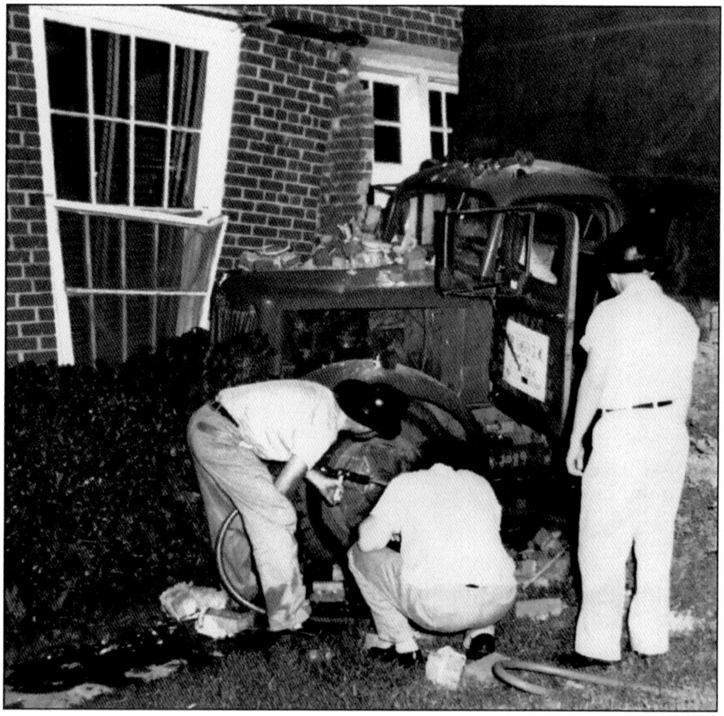

Truck Versus Building. Firefighters aim a high-pressure fog nozzle at a transfer truck that missed the roadway. Raleigh's first two American LaFrance 700 Series pumpers were high-pressure units. Delivered in 1951, each could deliver 800 PSI of pressure through the special nozzles. They were assigned to Engine 3 and Engine 5. (RFD.)

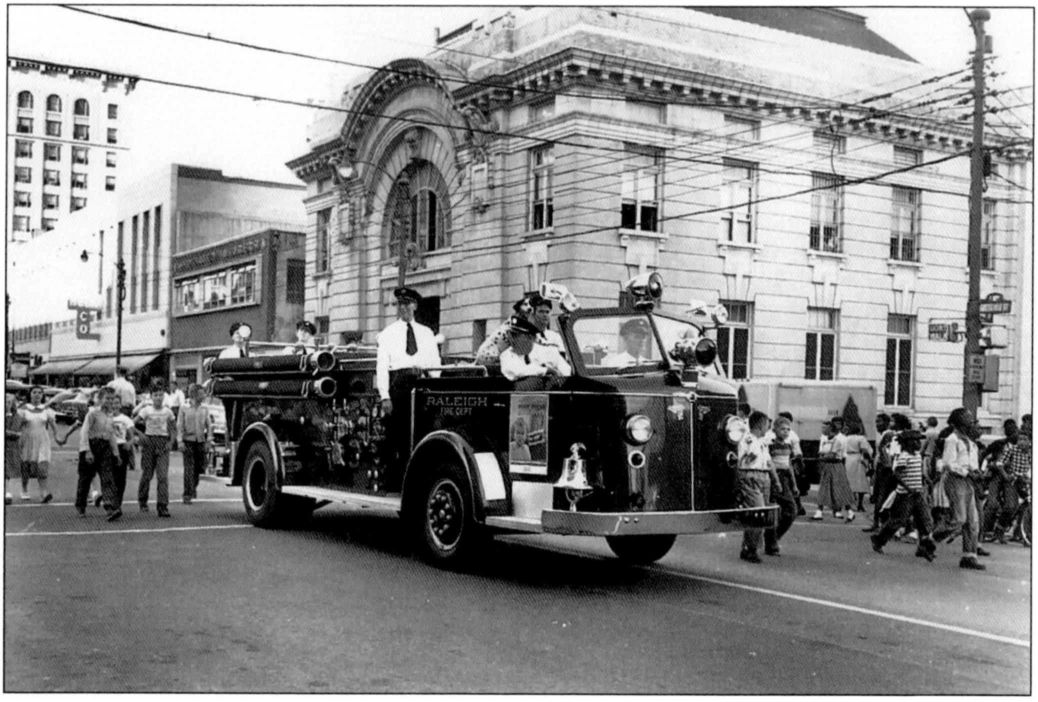

Fire Prevention Parade, October 1953. Riding past City Hall on Fayetteville Street are, from left to right, Raleigh firemen Capt. Ed Pollard, Capt. John Godwin, Capt. Lucius Q. "Luke" Godwin, and driver Floyd Pipkin. (*N&O* photograph.)

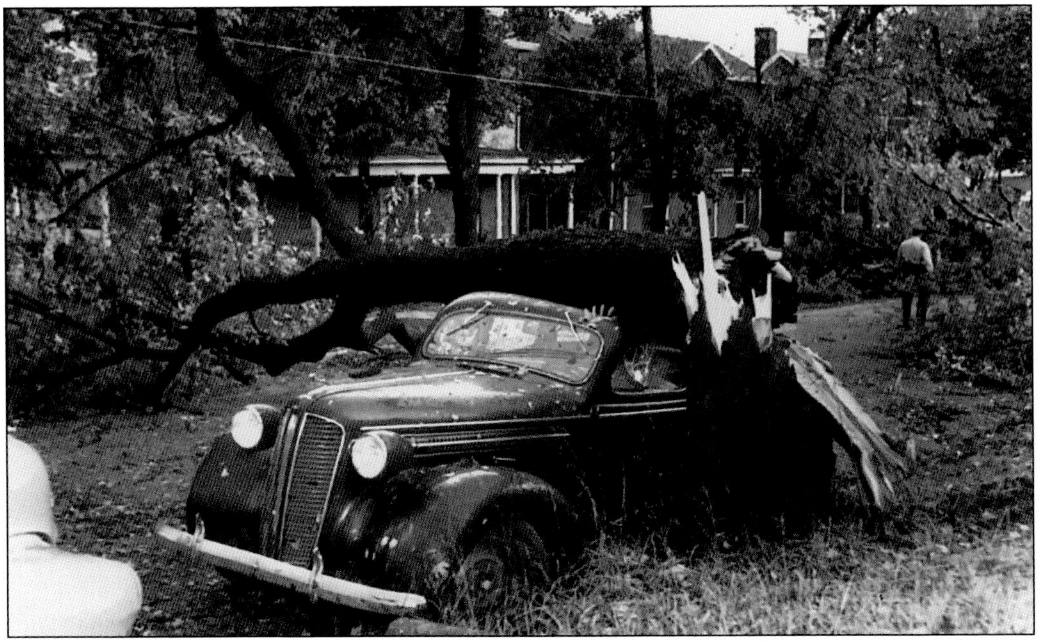

Hurricane Hazel, October 15, 1954. Making landfall in Brunswick County with the state's greatest storm surge to date, the eighth Atlantic hurricane of 1954 passed just east of the capital city. At Raleigh-Durham Airport, 90-mph gusts were recorded around 1:30 p.m. The Raleigh Fire Department answered 15 alarms that Friday, ranging from "electric wires" to "tree on fire." (*N&O* photograph.)

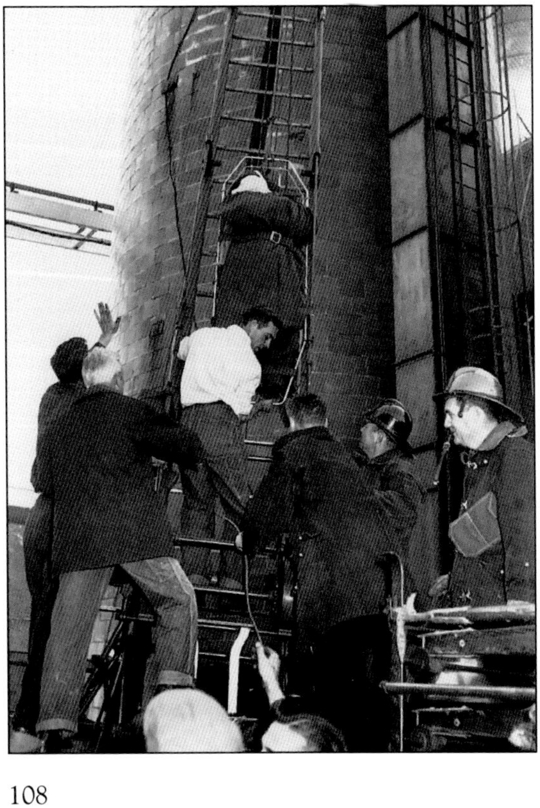

Rescued Worker, February 18, 1957. Electrician and ministerial student Winfred L. Welborn is freed after being buried to his chin for four hours in a coal elevator at the Wake Finishing Plant on U.S. 1, about eight miles north of Raleigh. He was discovered about 7:15 a.m. with a safety rope still around his chest. He had entered the 25-foot diameter elevator to unclog the coal, but his rope was too long and he became stuck. The touchy rescue operation included building a square wall to protect his head. Dr. C.T. Wilkerson of Wake Forest also administered "hyperdermics" to the trapped victim to relieve pain and treat for shock. After a hole was cut in the side of the elevator and the surrounding coal removed by hand, Welborn was freed about 11:15 a.m. He was transported to Rex Hospital. (*N&O* photograph.)

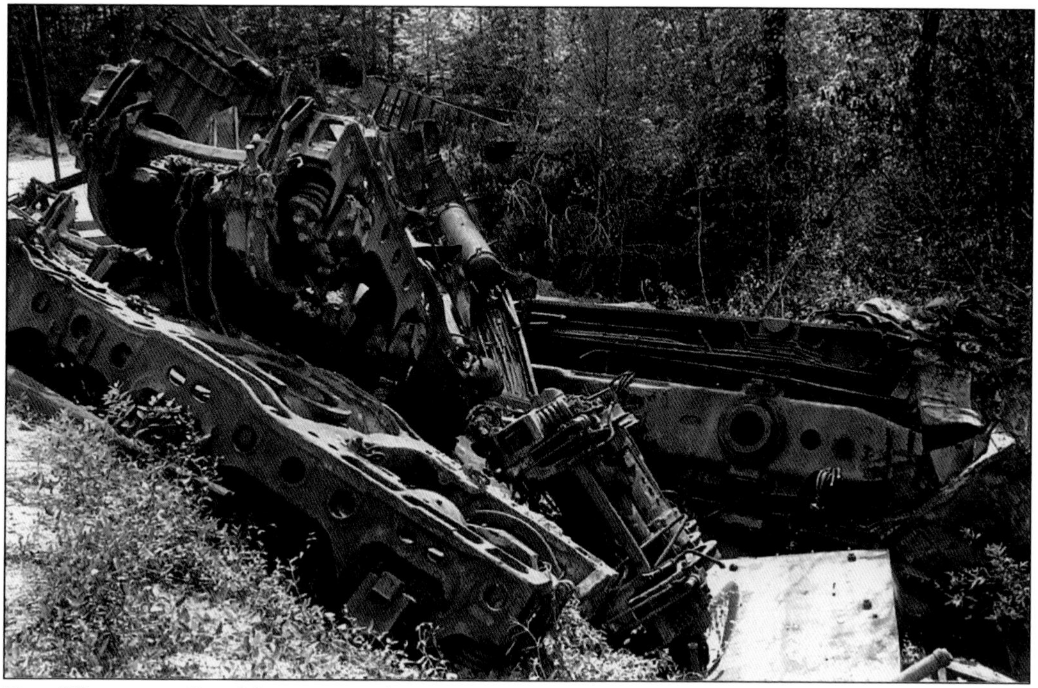

The Wreck of Old Number 4, August 1963. One month after plunging off a trestle two miles east of Knightdale, the remains of a Norfolk and Southern freight train rusts in Mark's Creek. Three trainmen were killed in the July 1 accident after lightning burned the wooden bridge. One of the victims, an apprentice making his first run, was pinned under a diesel engine for over an hour. (*RT* photograph by Lawrence Wofford.)

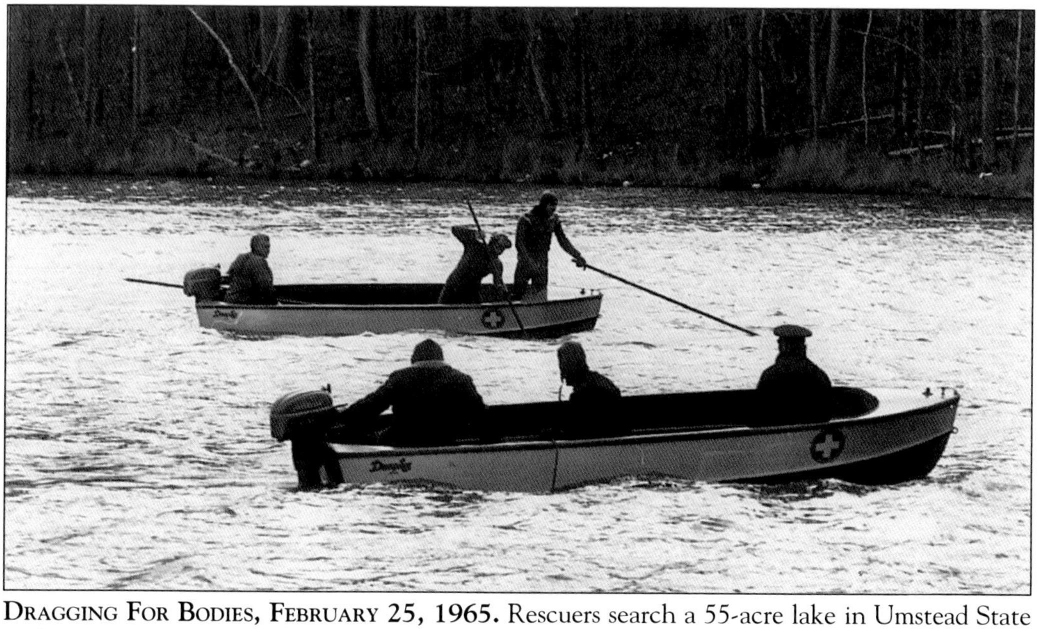

Dragging For Bodies, February 25, 1965. Rescuers search a 55-acre lake in Umstead State Park about two miles south of the Raleigh-Durham Airport after a light plane crashed into the freezing waters the night before. All five aboard the aircraft drowned. (*N&O* photograph.)

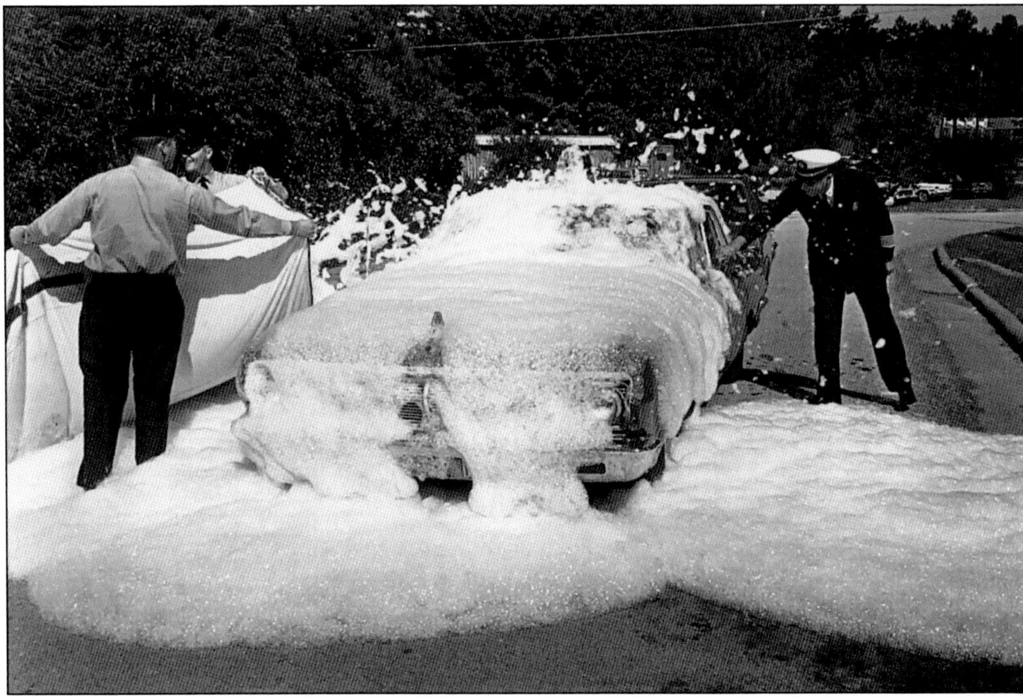

Greeting the Chief, October 1965. Checking on Durham fire chief C.L. Cox is his capital city counterpart Jack Keeter. The Durham dignitary was greeted at the city limits on Highway 70 by motorcycle policemen who escorted him to the Glenwood Shopping Center where the fire department's foam tuck was waiting. The cities were competing for lowest fire losses during Fire Prevention Week. (*RT* photograph by Ken Cooke.)

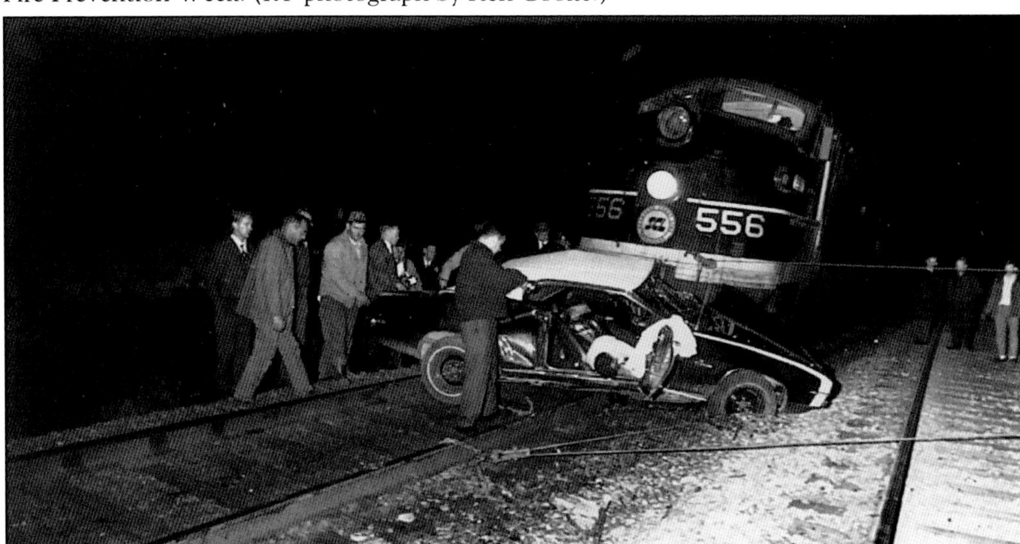

They Tried to Beat the Train, November 20, 1969. Two people were killed and a third jumped to safety when their automobile stalled at the Reedy Creek Road crossing east of Cary. The 17-year-old survivor said the warning lights were flashing when they attempted to cross. The 1968 Camaro with its 23-year-old driver and 17-year-old passenger was pushed about a quarter-mile by the southbound Seaboard Coast Line engine. (*RT* photograph by Hal Barker.)

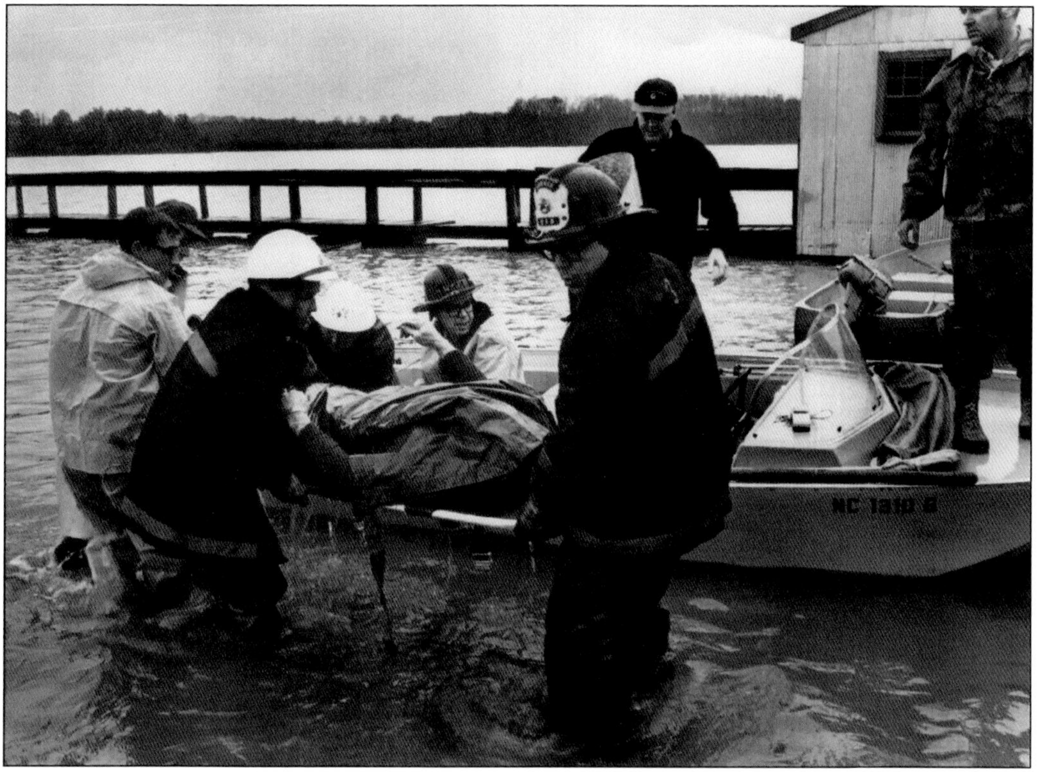

DROWNING, FEBRUARY 13, 1971. Emergency workers remove from Lake Benson the body of a Garner man. Personnel began searching the lake after his car was found in 12 feet of water near the bridge on Aversboro Road. He had been missing for three days. (*GLF* photograph.)

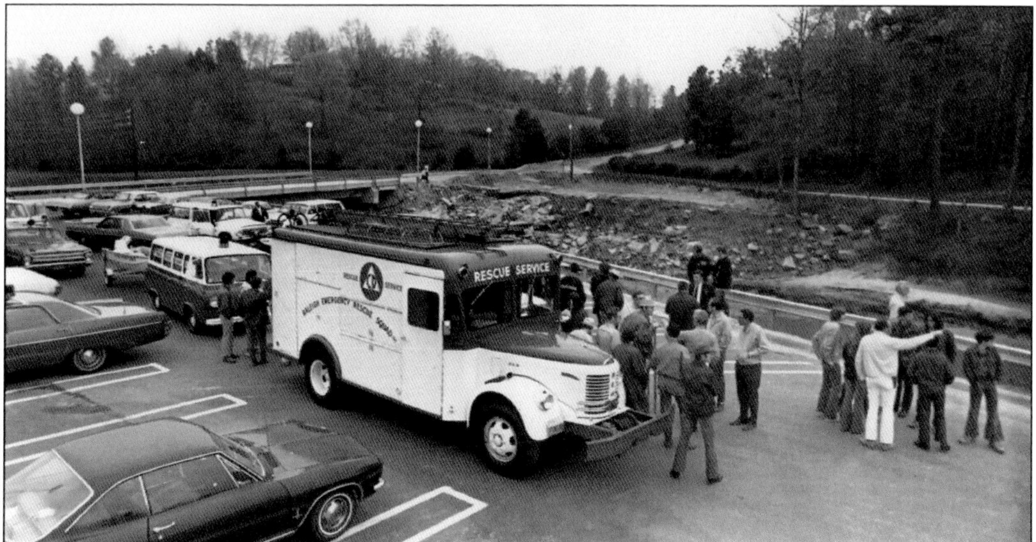

CRABTREE CREEK, APRIL 7, 1973. Rescue squad members and other volunteers search behind Crabtree Valley Mall for the body of a six-year-old boy. His family's canoe had overturned upstream in the rain-swollen creek several days earlier. About 200 searchers participated that day. (*N&O* photograph by Steve Adams.)

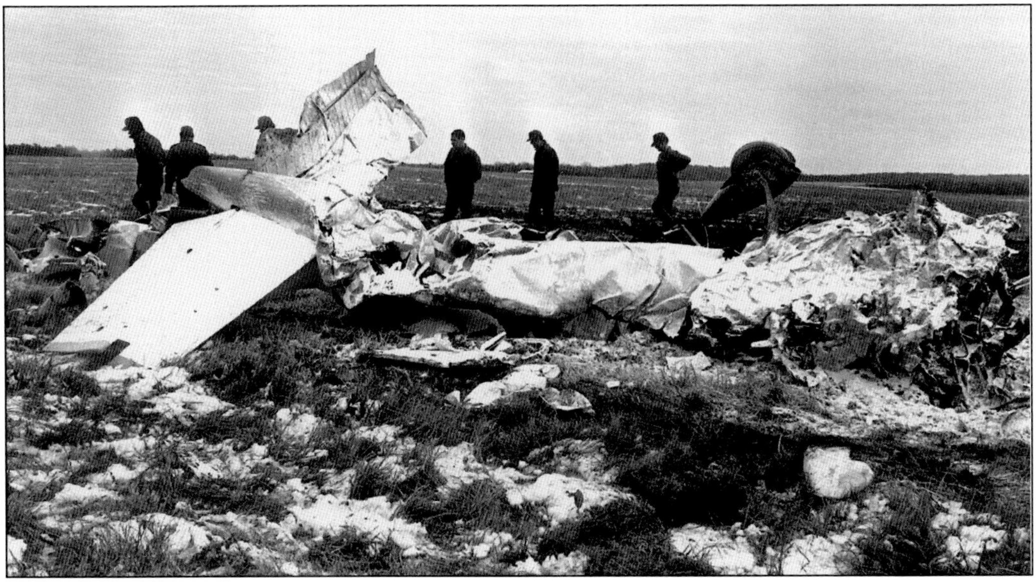

Midair Collision, December 4, 1971. Two people were killed aboard this private plane after colliding with a DC-9 about three miles southwest of Raleigh-Durham airport. Both aircraft were approaching for a landing when the single-engine Cessna Sky Wagon became lodged under the passenger jet. It subsequently dropped beside a runway and exploded in full view of air travelers and ground personnel. The airliner circled for an hour before landing safely on the foamed tarmac. (*N&O* photograph by Warren Uzzle.)

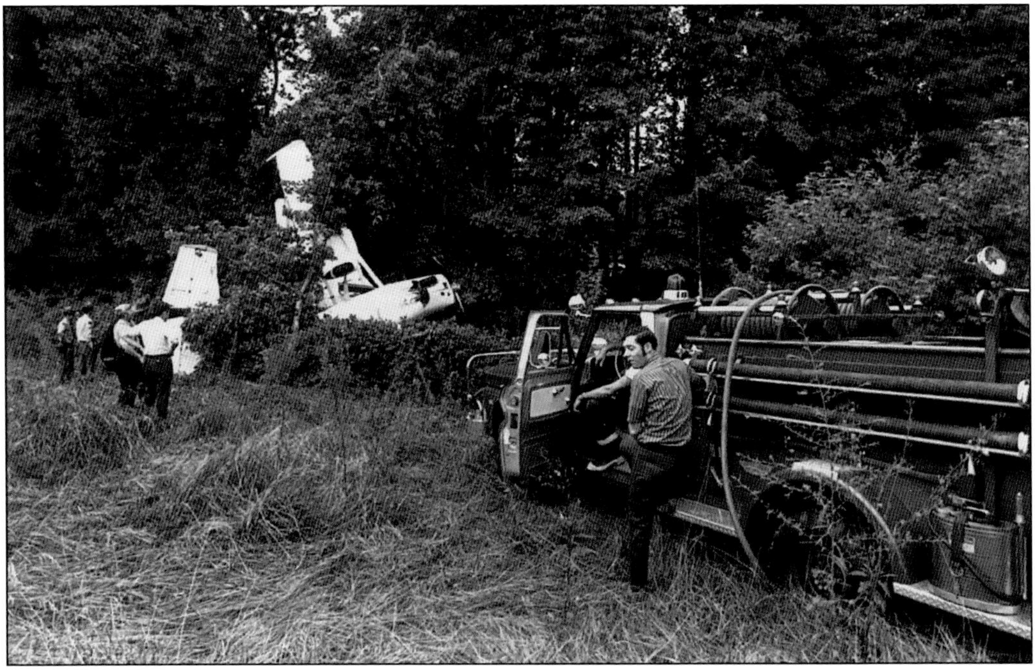

Safe But Sorry, July 28, 1973. Four people walked away unhurt when this Cessna 172 crashed near Melrose Drive just east of Raleigh. They had just taken off from a nearby airstrip when the aircraft failed to clear a group of trees. The New Hope Fire Department responded with two trucks. (*N&O* photograph by Steve Murray.)

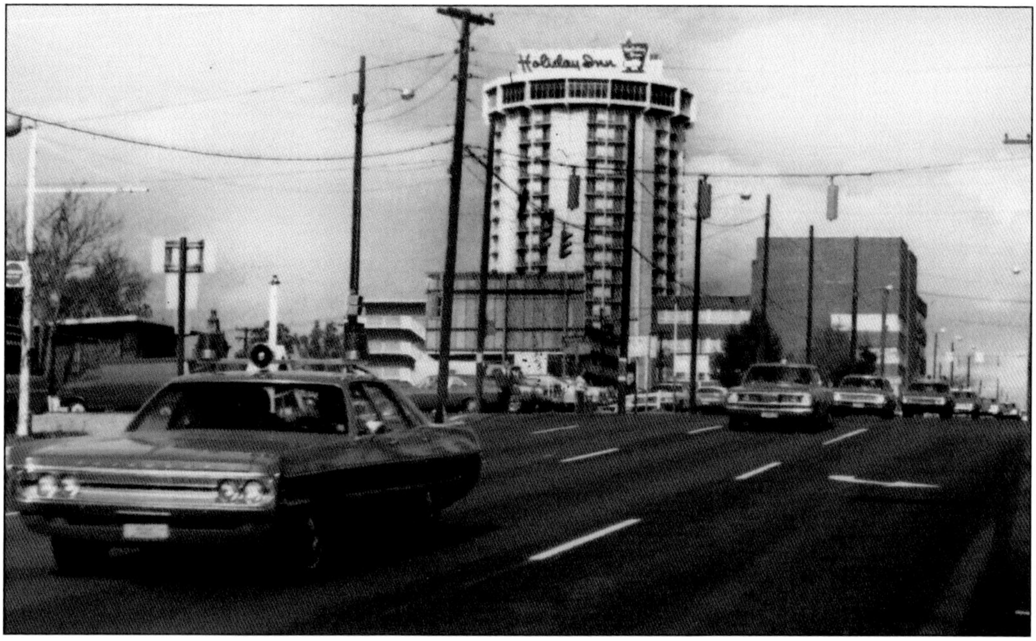

FUNERAL PROCESSION, NOVEMBER 12, 1974. Hundreds of firefighters, some from as far as Greensboro, honored Raleigh fire chief Clarence R. Puryear after he suffered a fatal heart attack at home. The 33-year veteran of the department had been chief just 17 months. He was buried at Montlawn Memorial Park. (RFD.)

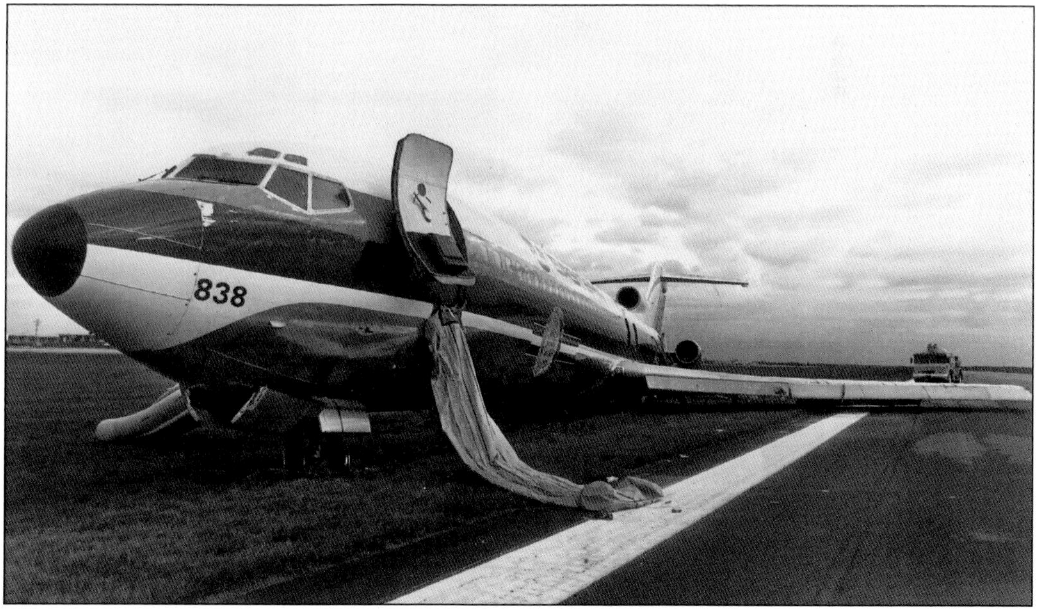

HARD LANDING, NOVEMBER 12, 1975. This Eastern Airlines jet crashed while landing in a heavy rainstorm at Raleigh-Durham Airport. The Boeing 727 struck the ground approximately 282 feet short of the runway about 8 p.m. Flaps were sheared off both wings, and the belly caved in after the landing gear collapsed. Four people were slightly injured in the rush to evacuate the craft. (*N&O* photograph.)

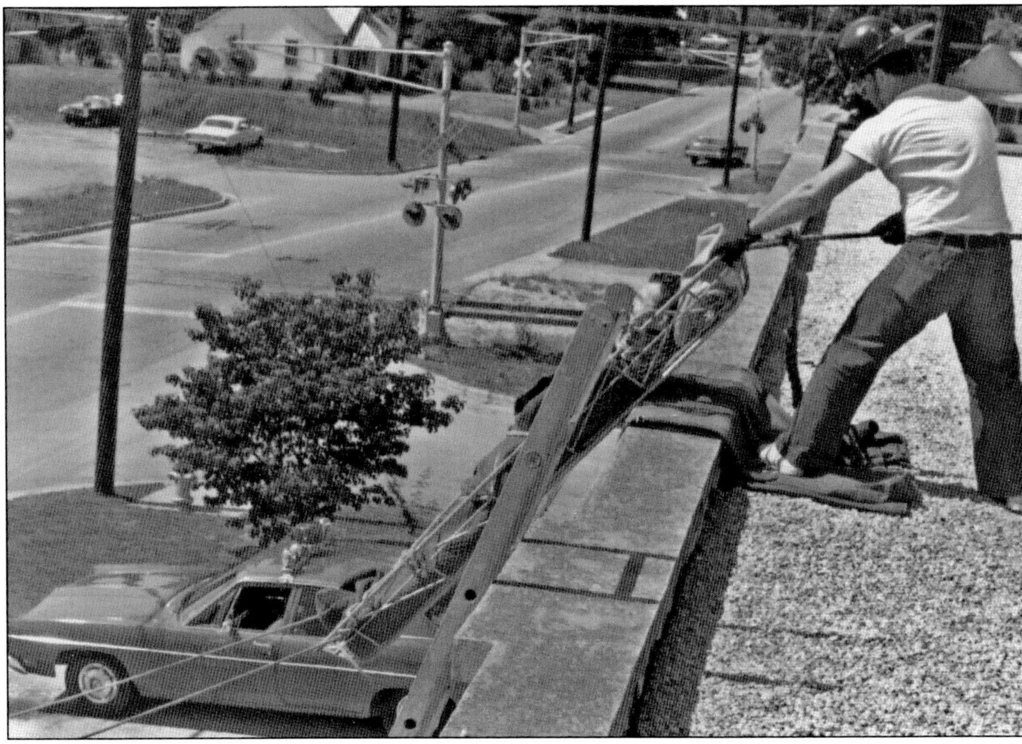

Rescue Training, c. 1975. Cary firefighters practice lowering an "injured" person from the roof of Station 1 at 100 N. Academy Street. (CFD.)

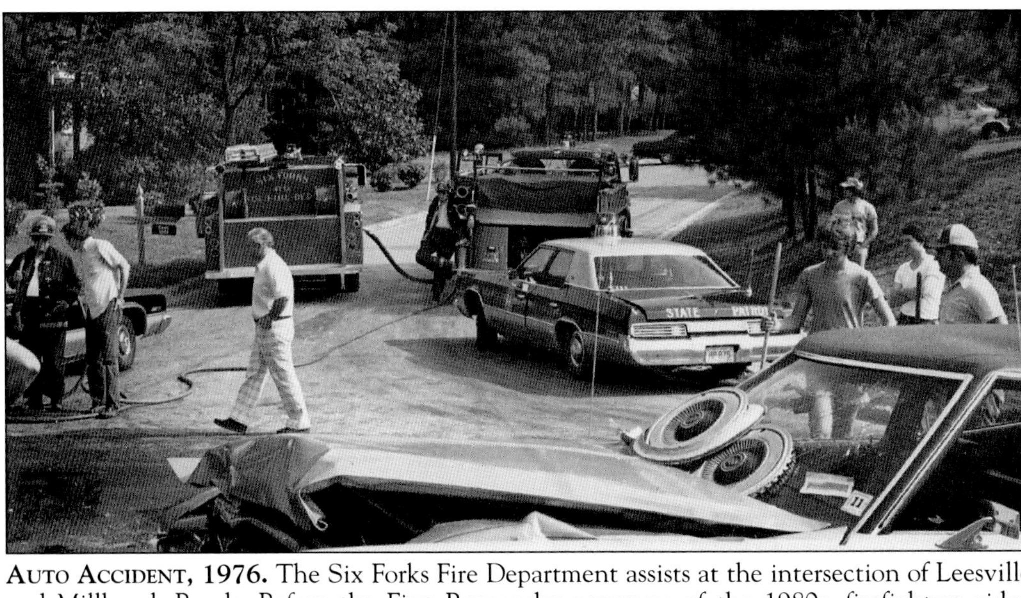

Auto Accident, 1976. The Six Forks Fire Department assists at the intersection of Leesville and Millbrook Roads. Before the First Responder programs of the 1980s, firefighters aided at accident scenes with such duties as directing traffic, washing away debris, and, of course, protecting against fires. (Don Adams photograph.)

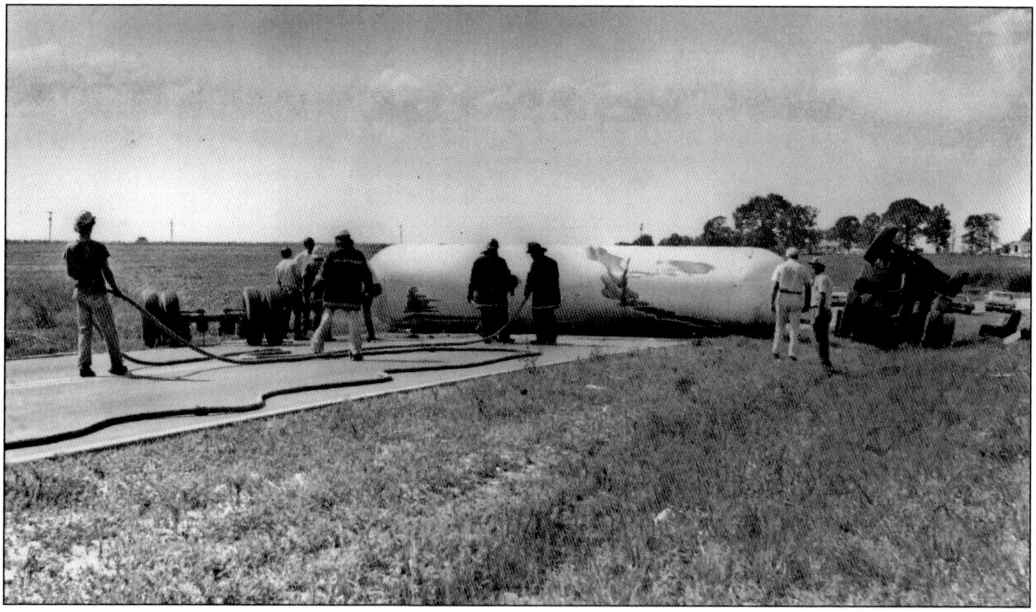

LIQUID PETROLEUM, AUGUST 23, 1976. The driver of this LP gas truck was killed after it overturned three times on Highway 55 south of Fuquay-Varina. Landing across both lanes of traffic, other drivers immediately spotted the cloud of leaking propane and began backing away. By grim coincidence, the victim's father witnessed the wreck and was first on the scene. Units from both Fuquay-Varina and Angier-Black River responded. (FVFD.)

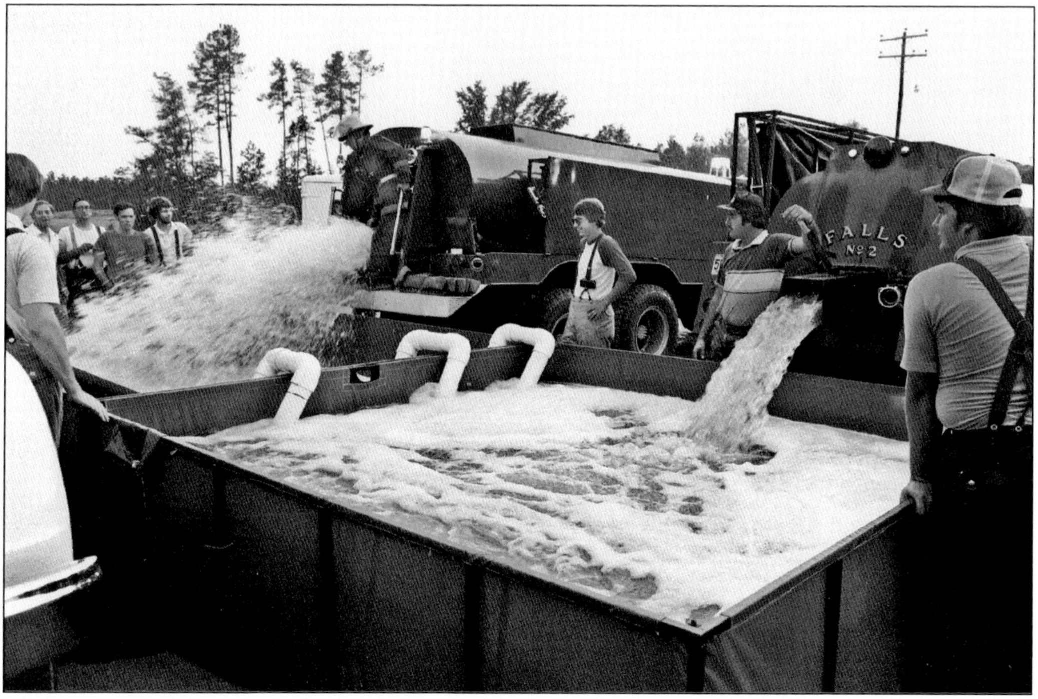

WATER SUPPLY SCHOOL, AUGUST 1981. Tankers from Stony Hill (left) and Falls (right) participate in a training exercise in Wake Forest. (*WW* photograph by Bob Allen.)

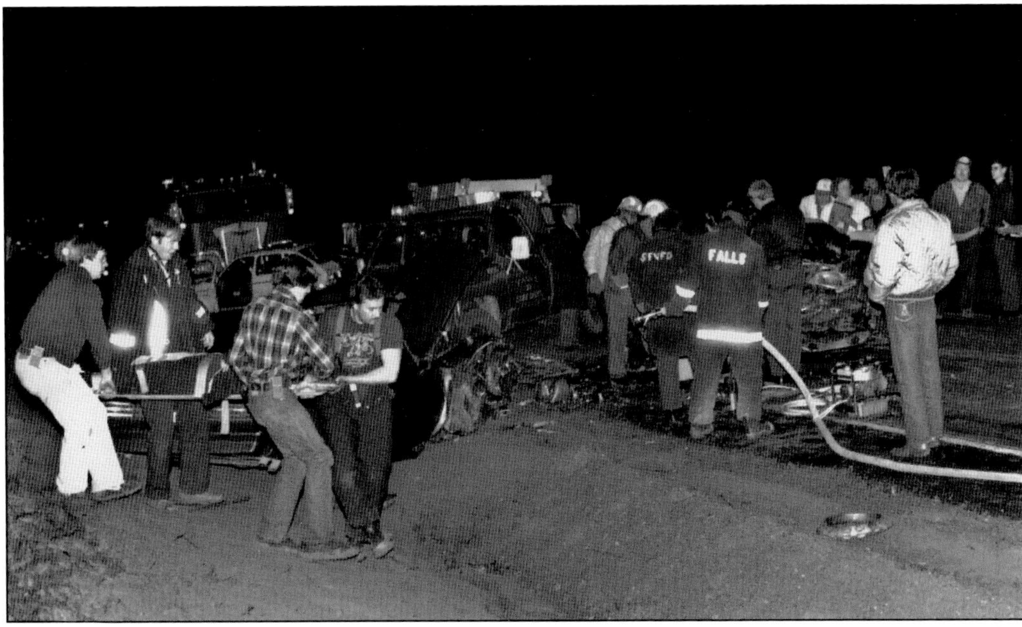

BAD WRECK, MARCH 6, 1984. This non-fatal collision on Durant Road about a mile-and-a-half west of U.S. 1 was the first First Responder call answered by the Falls Fire Department. They were the second fire department in the county to join the program after Fairview's pilot participation in March 1982. (FFD.)

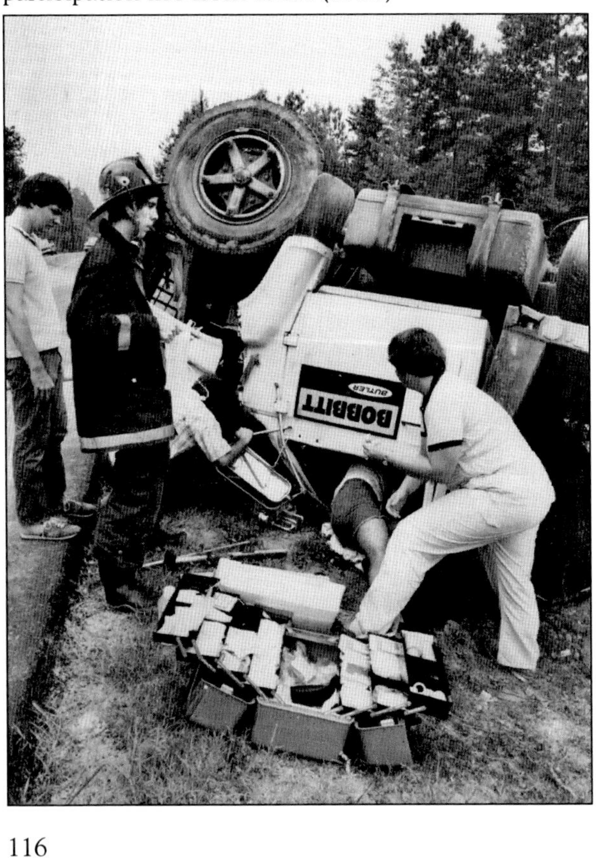

SUBJECT PINNED, AUGUST 14, 1984. Trapped for 30 minutes, the driver of this overturned truck on Durant Road was extricated and treated by members of the Falls Fire Department, New Hope Fire Department, and Northern Wake Rescue Squad. Rescuers used a Hurst tool to partially lift the truck while others administered first aid. Two arriving wreckers were promptly pressed into service so the driver's door could be removed. The victim was transported to Wake Medical Center but not hospitalized. (FFD.)

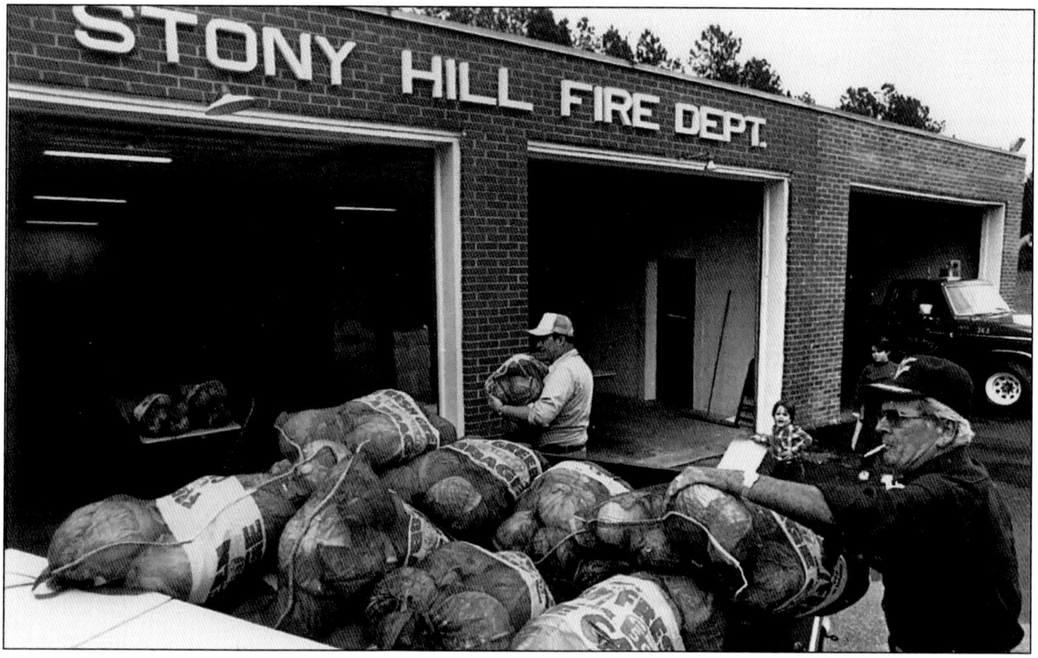

Texas Barbeque, November 1, 1985. Stony Hill fire chief Jimmy Perrott (left) and Tom Morris (right) unload cabbages for their Texas-style beef and chicken barbecue. Started in 1973 when a former chief donated a steer, the annual-turned-semiannual event has become the volunteer fire department's leading fund-raiser. That first year, they cooked 240 pounds of beef. This time, they're preparing 2,500 pounds of beef along with 800 chickens, 2,000 pounds of potatoes, and 1,500 pounds of cabbage for coleslaw. The Saturday event drew 2,940 people and netted $8,716. (*N&O* photograph by Robert Willett.)

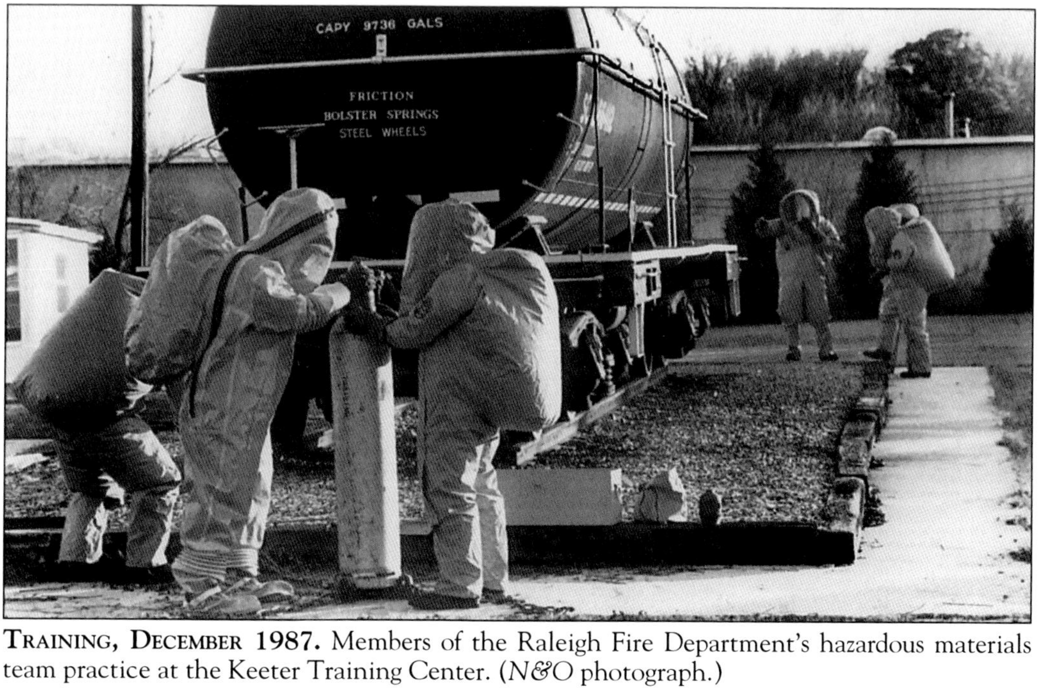

Training, December 1987. Members of the Raleigh Fire Department's hazardous materials team practice at the Keeter Training Center. (*N&O* photograph.)

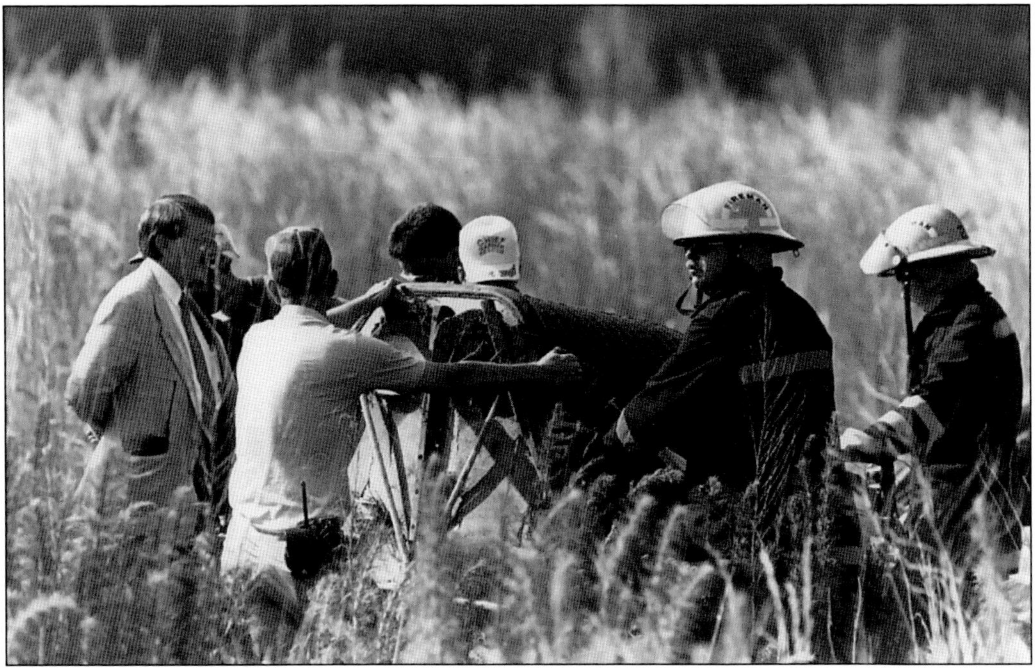

BALLOON ACCIDENT, JUNE 11, 1989. Three people were killed during a Sunday birthday ride when their hot-air balloon struck the wires of a 2,000-foot television tower off Highway 70 East, just inside the Wake County line. (*N&O* photograph by Gary Allen.)

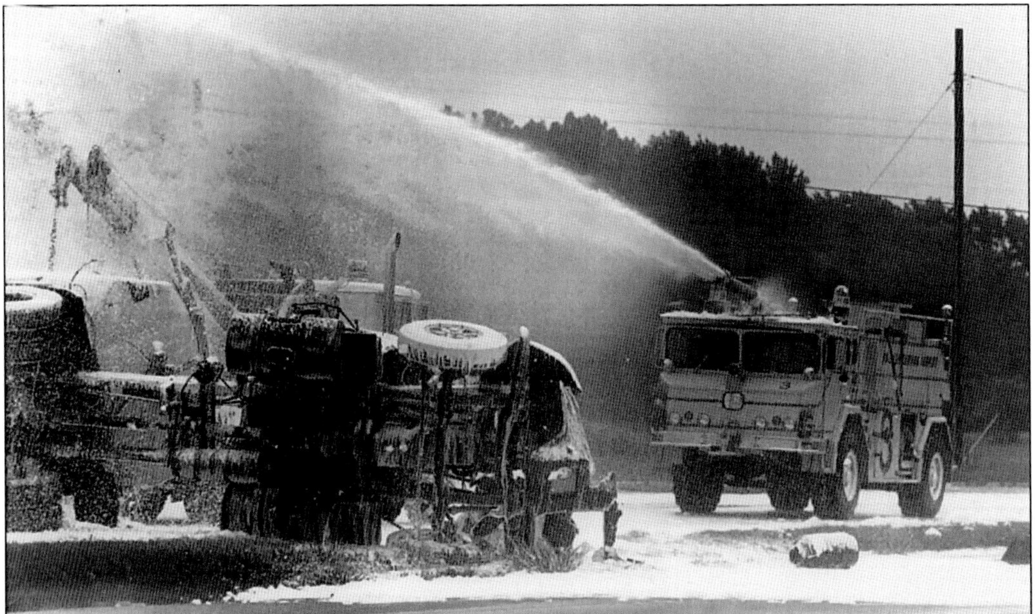

FLAMMABLE, JULY 31, 1989. This 8,000-gallon gasoline tanker collided with a car and overturned east of Garner on Highway 70 at Guy Road. The driver of the car was killed and over 1,000 gallons of gas were spilled. Thirteen fire departments from Wake and Johnston Counties responded, including this crash truck from Raleigh-Durham International Airport. The highway was closed for several hours. (*N&O* photograph by Scott Sharpe.)

Six

ARTIFACTS

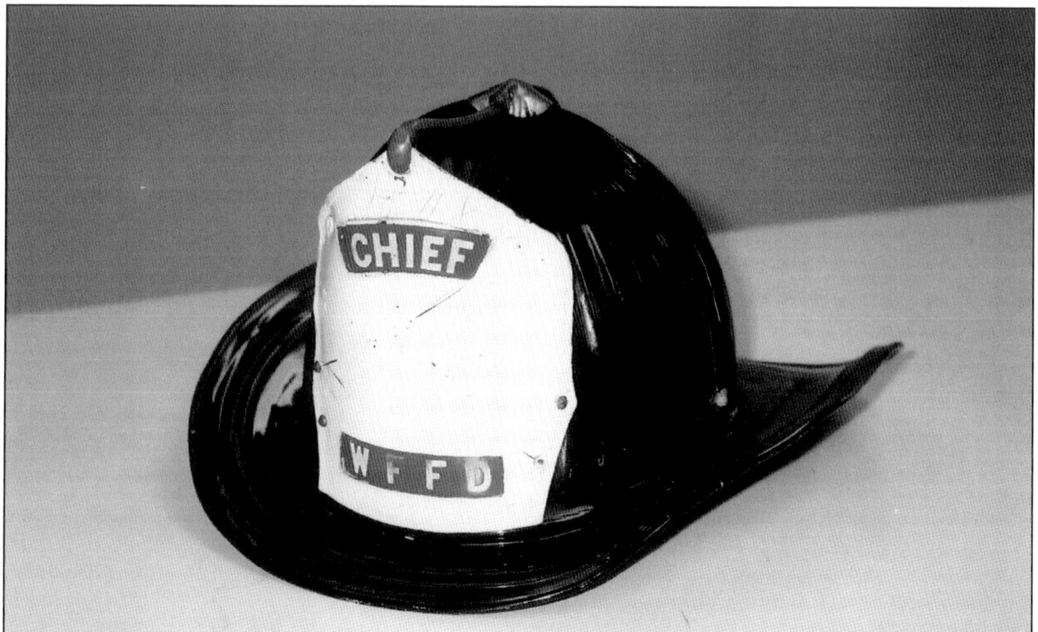

HELMET. This metal Cairns and Brothers helmet is displayed at Wake Forest fire station 1. (Mike Legeros photograph.)

FIRE ALARM BOXES

AND LOCATIONS

Wake Forest, N. C.

No. 12 Corner White & Spring Sts.
No. 13 Corner N. Main with Campus
No. 14 Corner Pine & N. College Sts.
No. 15 Corner N. Main & Walnut Sts.
No. 16 Corner N. Main & Briggs Sts.
No. 17 Corner E. Juniper & Lewis Streets
No. 18 Corner Spring & Davis Sts.
No. 21 Out of TownFires
No. 22 Corner S. Main & Dunn Sts.
No. 24 S. Wingate & W. Sycamore Streets
No. 31 Corner White & Waite Sts.
No. 32 Corner White & Owens St.
No. 33 Elm St. at Filter Plant
No. 34 S. Main between Elm & Vance Streets
No. 35 Corner S. Main & E .Sycamore Streets
No. 36 S. Main with Campus
No. 37 Corner West Ave. and Wingate Streets
No. 51 Corner Waite & Taylor Sts.
No. 52 W. Sycamore at Dorsett Sts.

Locate YOUR box on map (over)

EMERGENCY CALLS

POLICE and FIRE

Police 215-1
Ambulance 246-1

FOR FIRE

Day 247-1
Night 215-1

DOCTORS

Dr. G. C. Mackie ....
Dr. G. W. Corbin .... } Infirm. 352-1, Office 324-1

Dr. C. T. Wilkinson .... } Office 426-1, Home 265-6

Dr. R. W. Wilkinson .... } Office 274-1, Home 323-1

WAKE FOREST COLLEGE 1947 FOOTBALL SCHEDULE

Sept. 27—Georgetown at Wake Forest (Night game)
Oct. 4—Clemson .... at Clemson
Oct. 11—N. Car. .... at Chapel Hill
Oct. 18—Geo. Wash. at Wash., D. C.
Oct. 25—Duke .... at Wake Forest
Nov. 1—Wm. & Mary at Williamsburg
Nov. 8—Boston .... at Boston
Nov. 15—N. C. State .... at Raleigh
Nov. 21—Duquesne .... at W.-Salem (Night game)
Nov. 27—S. Car. .... at Charlotte

BOXES. This list of firebox locations also featured a map on the reverse side. (Courtesy of Wake Forest College Birthplace Society.)

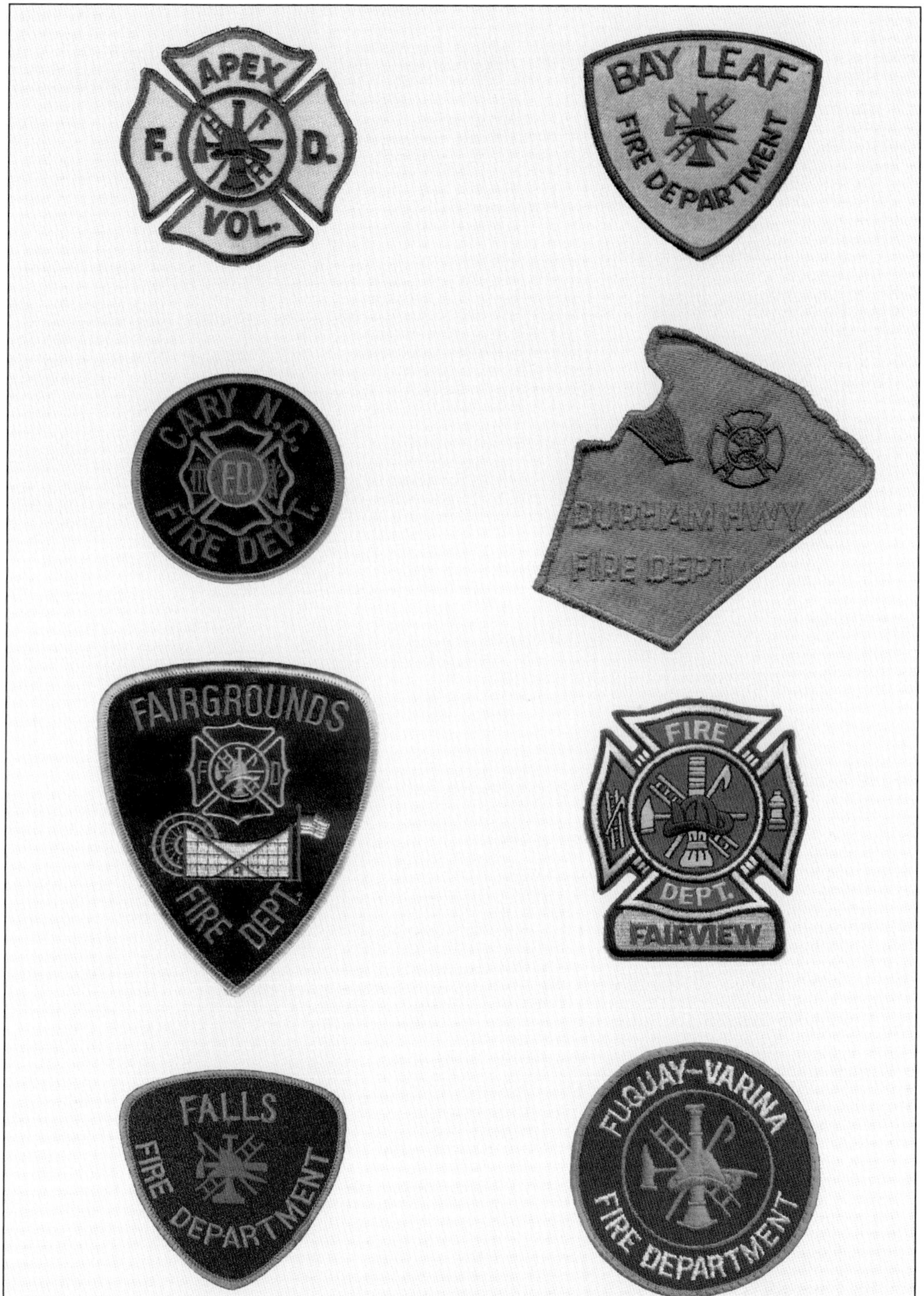

Patches. Pictured from left to right and top to bottom are patches from Apex, Bay Leaf, Cary, Durham Highway, Fairgrounds, Falls, and Fuquay-Varina fire departments.

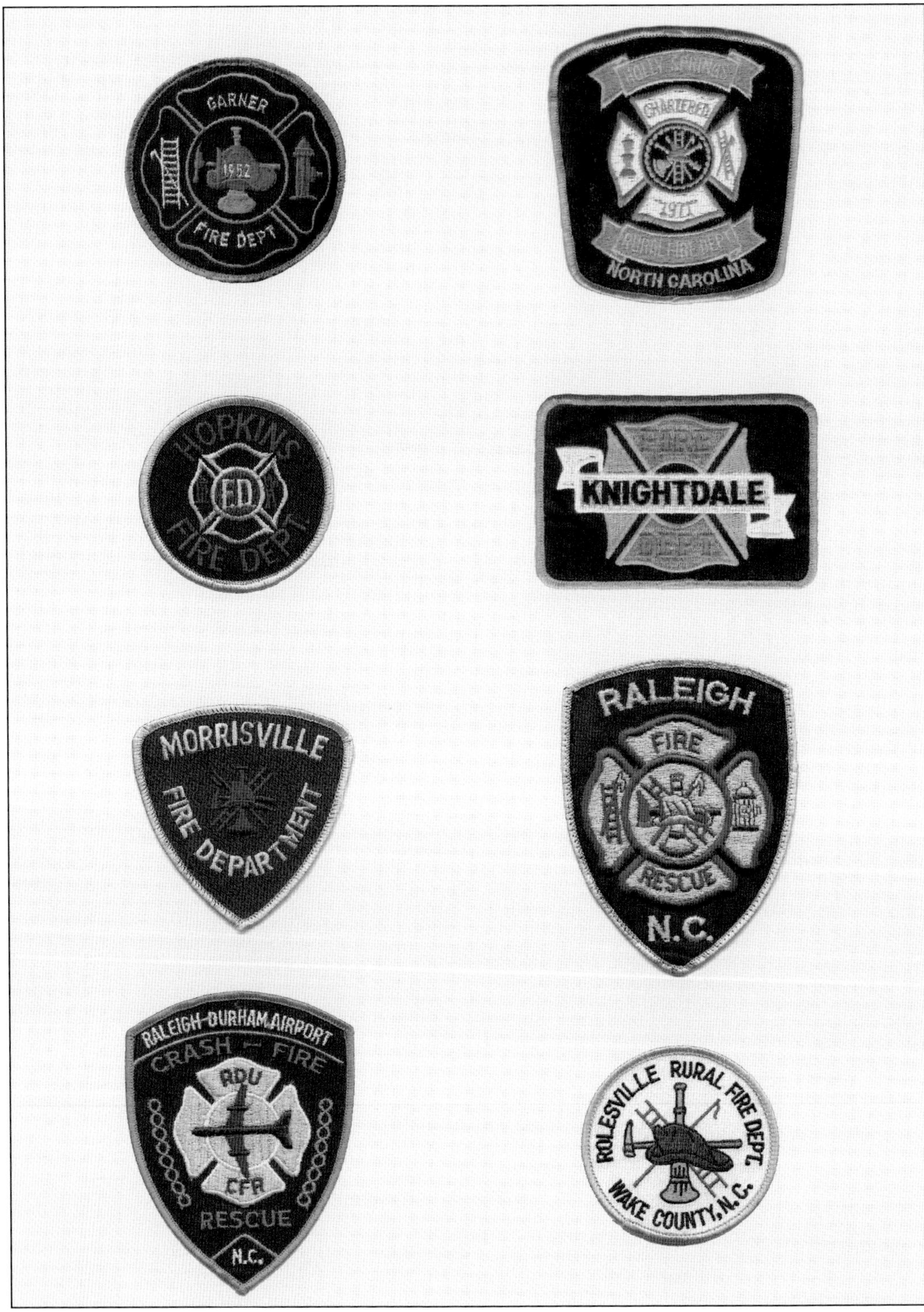

PATCHES (CONTINUED). Pictured from left to right and top to bottom are patches from Garner, Holly Springs, Hopkins, Knightdale, Morrisville, Raleigh, Raleigh-Durham Airport, and Rolesville fire departments.

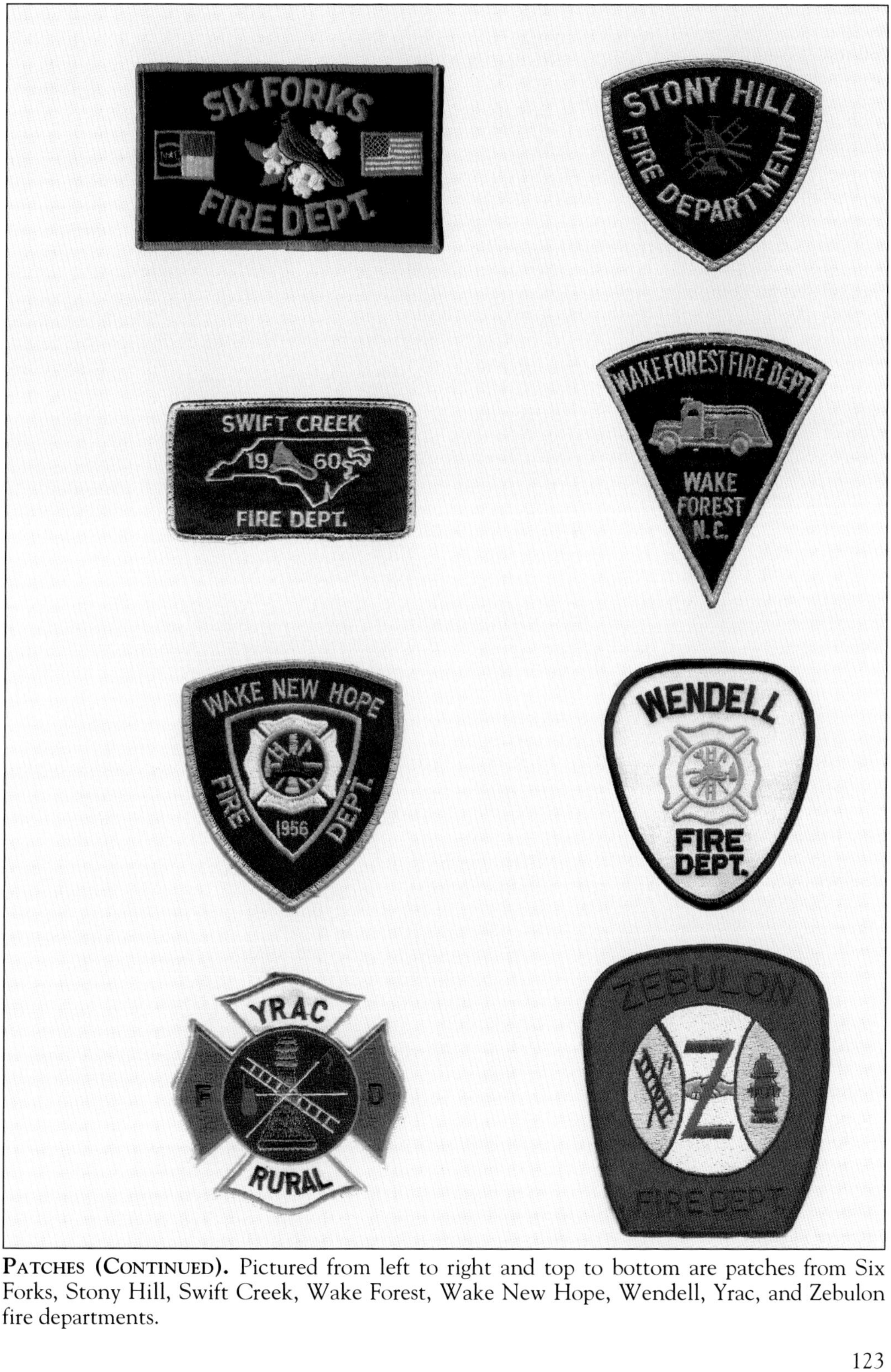

PATCHES (CONTINUED). Pictured from left to right and top to bottom are patches from Six Forks, Stony Hill, Swift Creek, Wake Forest, Wake New Hope, Wendell, Yrac, and Zebulon fire departments.

REPORT OF BOARD OF DIRECTORS AND SECRETARY

January 28 - December 31, 1957

The Western Boulevard Rural Fire Department, Inc., was chartered January 28, 1957. Ten members of the Western Boulevard Exchange Club signed the Charter and after the application was accepted by the Secretary of State, temporary officers were elected. It was decided to initiate a membership campaign on a basis of $10.00 for owners and $5.00 for renters.

After about six weeks it appeared there was sufficient interest to organize a fire fighting department and a public meeting was called on March 15, to which all interested citizens were invited. Approximately 150 people attended the meeting. By-Laws were adopted and, for the purpose of electing directors, the community was divided into four districts. Three directors were to be elected from each district at a subsequent meeting of the membership.

The second meeting was held on April 15, at which time directors were elected as follows:

1st DISTRICT		2nd DISTRICT	
Woodrow Cox	3 Years	John Allen	3 Years
W. M. Black	2 "	Lewis Bulwinkle	2 "
Alexander Goode	1 Year	Fred Shugart	1 Year
3rd DISTRICT		**4th DISTRICT**	
Dr. Henry Garren	3 Years	J. A. Holleman	3 Years
C. M. Godwin	2 "	Dale Graham	2 "
R. V. Liles	1 Year	I. O. Schaub	1 Year

Mr. R. V. Liles, finding his work would not permit his attending the meetings regularly, resigned and, as provided in the By-Laws, the remaining directors appointed Henry Garrison to fill the unexpired term of Mr. Liles.

The Board of Directors elected officers as follows:

Woodrow Cox, President
Dale Graham, Vice-President
I. O. Schaub, Secretary-Treasurer

Organizing and operating a corporation such as this is no easy task. Your Board has held eight meetings. All members have given hours of their time and some, even days.

One of the first actions of the Board was to appoint J. A. Holleman as Fire Chief and instructed him to investigate the kind of equipment needed and to secure prices on such equipment. It was soon learned that much of the main equipment needed was custom made and that usually, after an order was placed, it was three to four months before delivery could be expected.

The latter part of May a representative of the American Fire Apparatus Company of Battle Creek, Michigan, visited the community with a "demonstrator" pumper priced at $15,000. Since it had been used for a short time as a demonstrator, it was offered to your Board for $14,000.

Your Fire Chief and the Secretary visited five rural fire departments in Wake County and inspected their equipment, which cost less than the one offered us. However, not one of the engines we saw had as many desirable features for efficient fire fighting as was true of the outfit we were considering. Also, the Raleigh Fire Chief and other experienced fire fighters urged us to get good, modern, up to date equipment, but they did not recommend any specific manufacturer.

ANNUAL REPORT. The Western Boulevard Fire Department published this annual report in

Taking into consideration the possibility of perhaps a delay of four months or more and the many advantages of the outfit on the grounds, your Board voted unanimously to close a deal, paying $5,000 down and the remainder of $9,000 in four equal installments at 6% interest.

Your Board had hoped to get an outfit from Government surplus property but, on investigation, found that this was practically impossible. We were, however, fortunate in getting from surplus property two oil tank trucks for $125.00. By combining parts from the two and with donations from interested people and with hours of hard work by the volunteer firemen, you have a 2,000-gallon tank truck with a cash expenditure of less than $200.00.

We also secured from surplus property a "jeep" in good condition for $125.00. A portable gasoline mobile pump, for refilling the tank trucks, is being mounted on the jeep.

When the fire apparatus was secured, we needed a place for storage and a fire house centrally located. For a while, Mr. Pearson of the Esso Station kindly let the Department use his facilities. Mr. Swain of "Chicken House" fame offered the use of the old barbecue shed rent free for approximately eight years provided we would clean up and repair the building. This offer was accepted and again the volunteer firemen, with hours of hard work and generous donations of material from interested citizens and business concerns, have provided you with an insulated, heated fire house with a cash outlay of just a little over $200.00.

The Wake County Board of Commissioners offered to contribute $100.00 per month to rural fire departments that met certain minimum requirements. We qualified July 1, 1957, and since that date we have received each month $100.00 toward the cost of operation.

Also, we have been working to meet the minimum requirements of the N. C. Fire Insurance Rating Bureau so as to get a reduction in our insurance costs. It has taken longer than we had hoped but I am glad to report that we have had our inspection and that we more than met the minimum requirements. Unless something unforeseen develops, you may expect very shortly a reduction of 60¢ per $1,000 of insurance on dwellings and a greater percentage reduction on business establishments.

Your volunteer firemen are giving not only their time but risking injury and perhaps their lives in fighting fires. For their protection, your Board carries accident and hospital insurance. Also, fire, theft and liability insurance is carried on your equipment. It is costing over $300 per year for this protection but it is necessary.

There are approximately 1,500 homes and business establishments in our approved area. Up to December 31, 1957, 596 members had joined and paid their membership dues. They paid in $6,405.00. Unfortunately, we have many citizens who are not able to pay their dues but they deserve any protection we can give. There are, however, probably 500 or more who are able to pay who will get the insurance reduction and who will call for help if they have a fire but who have not joined the organization. Perhaps many of these are waiting for someone to invite them to join. Most solicitation has been done by a comparatively few people. Is it too much to expect all members to actively solicit their neighbors? If we had 1,000 paying members, our necessary annual assessments would be very materially reduced.

1958 after their first year of operation. (WWFD.)

Home Fire Protection Manual

• WHAT TO DO WHEN FIRE STRIKES
• HOW TO PREVENT FIRE

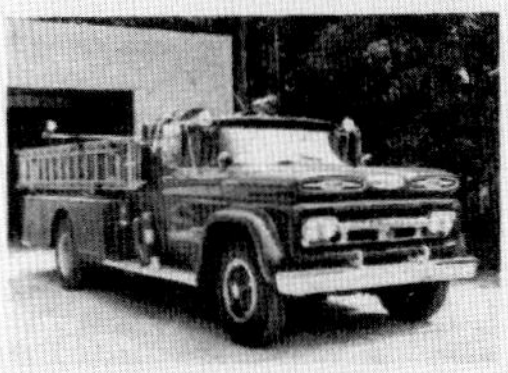

MODERN 500 Gal Pumper
Cost $20,000

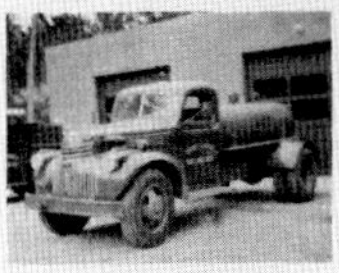

Auxiliary Tanker

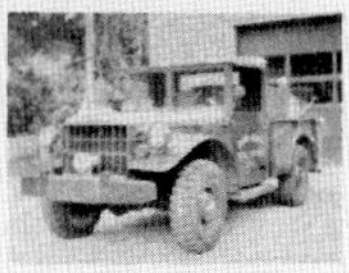

Brush and Field Fire Truck

SWIFT CREEK RURAL
FIRE DEPARTMENT, INC.
TO REPORT A FIRE PHONE TE 2-4400

THIS BOOK MADE POSSIBLE BY THE ADVERTISERS
PATRONIZE THESE CIVIC MINDED BUSINESS PEOPLE

Fire Prevention Manual. Distributed in 1963, this 28-page booklet also contained a financial report, local ads, and this plea from Fire Chief Stanford Mizelle: "The amount of money that is now being received is NOT ADEQUATE for the support of the department. We have heretofore asked that each family contribute just $10.00 per year to the support of the department. This amount is less than the sum that has already been saved by each family in the district on fire insurance rate reductions. If each family will pay this small annual sum we will be able to maintain the department as it should be. However, last year some families paid nothing towards the support of the department, but they derived the same protection and reduced insurance rates as those families that did. If you have not paid your fee for this year, please give it to one of the firemen or mail it. Please don't ask or expect that the Firemen, in addition to their service in protecting you, have to go out and practically beg for the money to maintain the Department." (SCFD.)

Garner Volunteer Fire Department
PEARL STREET
GARNER, NORTH CAROLINA
Serving the Town of Garner and Saint Mary's Fire District

☆ ★ ☆

DONATION DAYS
OCTOBER 8th AND 15th

Help Us To Help You

FINANCIAL STATEMENT & REPORT ENCLOSED

Amount .... Donated by:

Name ....

Address ....

Please, place your donation in this envelope, seal, fill in your name and address and have ready for your fireman when he calls October 8, outside city, October 15, inside city. If you are not at home on the first visit, please phone EM 2-9120 donation day or any Tuesday night and we will call back. Your donation may be mailed to the above address.

A receipt will be mailed to you upon request.

In Case of Fire, Dial VA 8-3459

SEE "LIFE SAVING" MESSAGE ON BACK

"You Can't Beat Fire Prevention to Save Your Life"

Donation Envelope. Used by the Garner Fire Department in the 1960s, this envelope also listed "good rules for fire safety." "Don't use gasoline for any purpose in the home. One gallon has the explosive power of 80 sticks of dynamite!" "Have heating units checked each fall by a competent mechanic before starting." "Store matches out of the reach of children." (GFD.)

Cook Book. The Morrisville Fire Department opened a substation in the Carpenter community in 1975. Located on State Road 1624 just west of Highway 55, the three-bay station was built by firefighters and other members of the community. Early equipment included a 1951 Dodge brush truck, a 1963 Chevrolet/Darley Champion pumper, and a 1969 Chevrolet tanker. The station and its apparatus are also pictured on the cover of this cookbook from 1975. Recipes included Mildred Howard's "Congealed Salad," Hilda F. Carpenter's "Persimmon Pudding," Onnie Riggsbee's "Cola Cake," Mary Frances Ferrell's "Mayonnaise Biscuits," Gayle Mills's "Hamburger Goulash," Mary Hill's "Porcupine Meat Balls," and Cleo Meachum's "Corned Beef Casserole." (MRFC.)

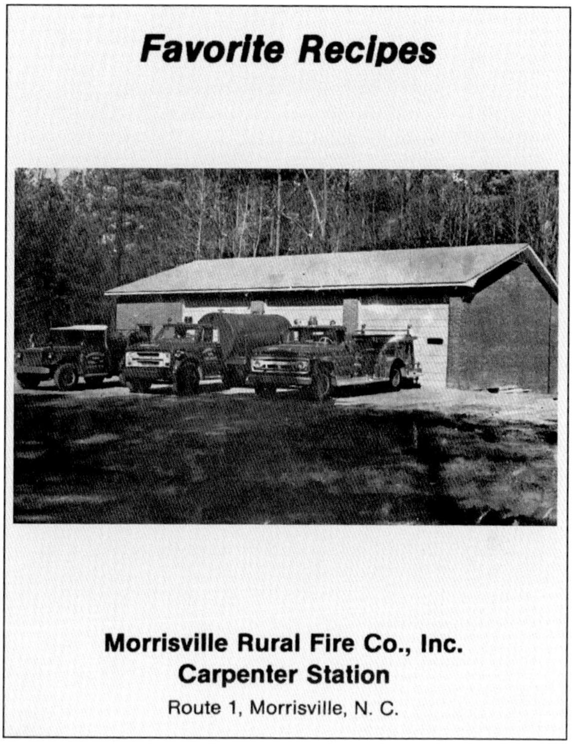

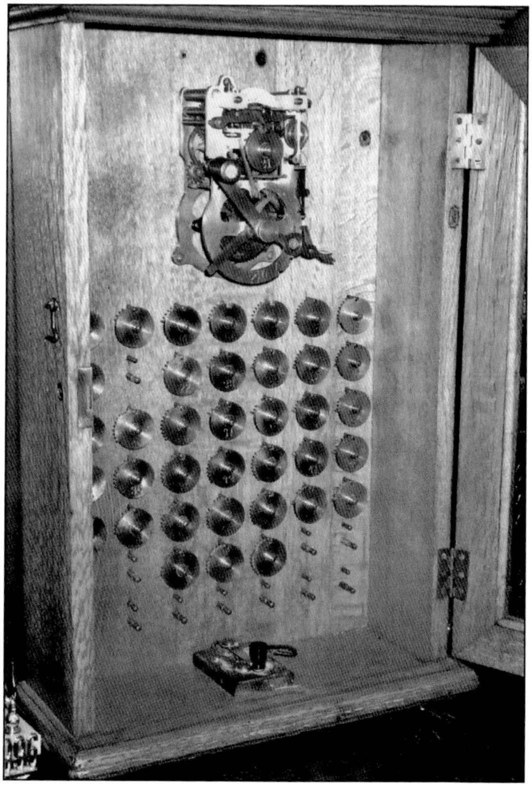

Manual Transmitter. This Gamewell manual transmitter was rescued from a trash can when Wake Forest's electric-telegraph fire alarm system was dismantled in the late 1970s. Installed in the early 1920s, the system tied the town's fireboxes to an air horn that announced box numbers. Volunteers carried wallet cards with the corresponding street addresses. For fires reported by telephone, the dispatcher selected one of the pictured metal discs. Each corresponded to a box number, and the dispatcher selected the one nearest the address of the emergency. For fires reported outside of the town limits, box 21 was used. Firefighters responded directly to the station for those alarms. (Mike Legeros photograph.)

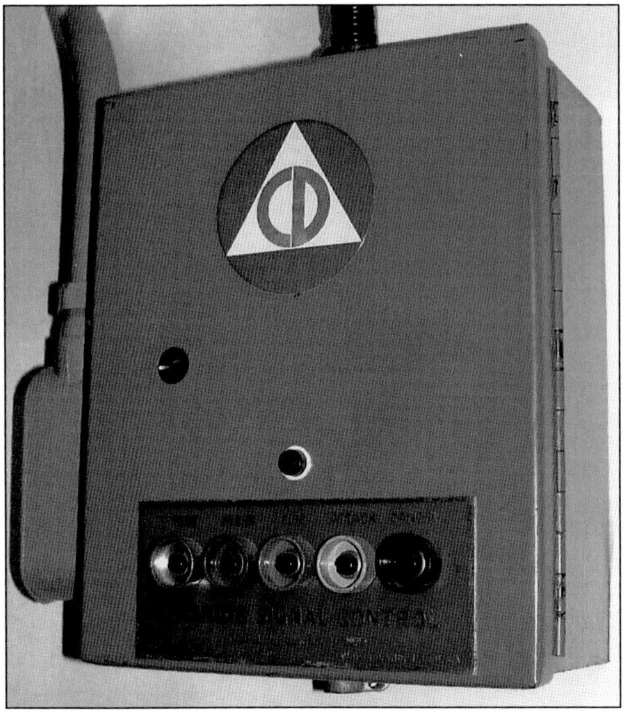

Control Panel. Used to alert volunteer firefighters, this SIRATROL timer control box at the Holly Springs fire station was connected to a two-tone Federal 3T22 Civil Defense siren outside. The control box produced three different sounds: a 3-minute wailing tone for an attack, a 3-minute steady tone for an alert, and a 3-minute high-low tone for a fire. Other Federal 3T22 sirens were located in Apex, Fuquay-Varina, Panther Branch (Garner Station 2), and Zebulon. (Mike Legeros photograph.)

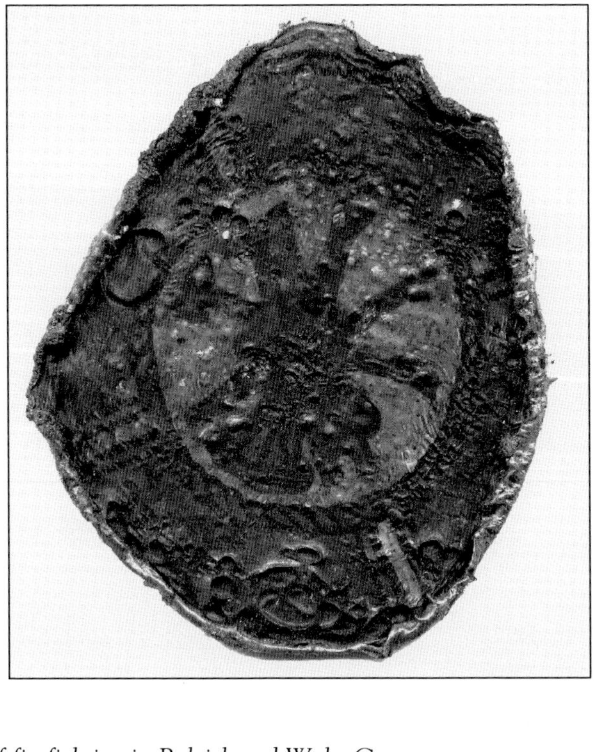

Melted. Pictured is the emblem from the author's helmet after fighting his first fire on January 28, 1990. He was asleep at Station 9 when the buzzer sounded at 3:53 a.m. "Structure fire." "6007 Applewood Lane." The run was a full four minutes. Two-story duplex. Fully involved. Engine 9 arrived first. The author backed the hose man. Flames blasted through the front door when they made entry. A body was later discovered. Smoke alarms had awakened two men upstairs. They jumped out a window, one breaking his leg. Their overnight guest, a 37-year-old female, was still inside. They tried to re-enter but were overcome by heat and smoke. The cause of the fire was not immediately known.

Help preserve the heritage of firefighting in Raleigh and Wake County. Contact mike@legeros.com for more information.